GEOLOGY UNDERFOOT
in ILLINOIS

Raymond Wiggers

1997
Mountain Press Publishing Company
Missoula, Montana

Fourth Printing, April 2009

All photographs by Raymond Wiggers
Cover painting © 1997 William Crook, Jr.
Bell Smith Springs, Shawnee National Forest

Many of the illustrations in this book are reproduced or modified
with permission from the Illinois State Geological Survey.

Excerpts from "The Windy City" in *Slabs of the Sunburnt West*
by Carl Sandburg, copyright 1922 by Harcourt Brace & Company and renewed 1950 by Carl Sandburg, reprinted by permission of the publisher.

Excerpts from "Prairie" in *Cornhuskers* by Carl Sandburg, copyright 1918
by Holt, Rinehart and Winston, Inc. and renewed 1946 by Carl Sandburg,
reprinted by permission of Harcourt Brace & Company.

From *The Way of Life* by Lao Tzu, translated by Raymond B. Blakney.
Translation copyright © 1955 by Raymond B. Blakney, renewed © 1983
by Charles Philip Blakney. Used by permission of Dutton Signet,
a division of Penguin Books USA Inc.

Lines from "Mrs. Sibley" and "Serepta Mason" from *Spoon River Anthology*
by Edgar Lee Masters. Originally published by the Macmillan Company. Permission by Ellen C. Masters.

Excerpt from "Plains" in *Collected Poems* by W. H. Auden, copyright 1976
by Random House, Inc., reprinted by permission of the publisher.

Library of Congress Cataloging-in-Publication
Wiggers, Ray.
  Geology underfoot in Illinois / Raymond Wiggers.
    p.   cm.
  Includes bibliographical references (p. -) and index.
  ISBN 0-87842-346-X (pbk. : alk. paper)
  1. Geology—Illinois. I. Title.
QE105.W6 1997
557.73—dc20
                                                              96-44732
                                                                 CIP

MP Mountain Press
PUBLISHING COMPANY
P.O. Box 2399 • Missoula, MT 59806 • 406-728-1900
800-234-5308 • info@mtnpress.com
www.mountain-press.com

CONTENTS

Preface   *ix*

Acknowledgments   *xi*

Illinois Back When: An Overview   *1*

Tips for Exploring the Sites Described in This Book   *11*

1 Lead Mines and Lovely Hills
Galena and Environs   *13*

2 A Diversionary Tale
Apple River Canyon State Park   *21*

3 A Scenic Bluff, a Zone of Faults, and the Track of a Mighty River
Mississippi Palisades State Park   *27*

4 The Upper Rock River: Part One
Lowden State Park   *33*

5 The Upper Rock River: Part Two
Castle Rock State Park to Dixon   *39*

6 A World of Wind and Melting Glaciers
The Princeton Area and the Green River Lowland   *45*

7 Cliffs, Canyons, and Catastrophes: The Upper Illinois River
Buffalo Rock and Starved Rock State Parks   *51*

8 A Land Like the Rolling Sea
Glacial Park and Environs   *59*

9 A Pocket History of a Freshwater Sea
Illinois Beach State Park   *67*

10 A Primer of North Shore Geology
Lake Bluff to Evanston   *75*

11 In the Land of the Fox
Batavia and Environs   *85*

12 The Architectural Geology of Downtown Chicago: A Sampler
The Near North Side and the Loop   *93*

13 Exploring the Lower Des Plaines
Lockport and Lemont   *107*

14 Blue Island Bracketed
Chicago's South Side   *113*

15 A Deep Look at the Distant Past
Thornton Quarry   *119*

16 In the Path of the Kankakee Torrent
Kankakee River State Park and Environs   *127*

17 The Realm of Fossil Forests
The Mazon Creek Area and the Longwall District   *135*

18 Across a Lazy River
Henry and Environs   *141*

vii

19 A Landscape Throws off Its Glacial Burden
    The Galesburg Plain   *145*

20 Homage to the Mississippian
    Nauvoo to Warsaw   *149*

21 The Havana Lowland
    Pekin to Havana   *157*

22 Reconstructing a Coal Age Stream
    North of Ripley   *165*

23 A Mystery from the Age of the Dinosaurs
    The Beverly-Baylis Upland   *171*

24 The 120-Million-Year Hike
    Pere Marquette State Park   *177*

25 Moraine-Surfing Down the Interstate
    Pontiac to Le Roy   *183*

26 A Tale of Two Rivers
    Monticello and Environs   *191*

27 Coal Country and Pennsylvanian Cyclothems
    Forest Glen County Preserve and Environs   *197*

28 On the Edge of Two Ages
    Shelbyville and the Interlobate Complex   *203*

29 The Black Bounty of the Illinois Basin
    Robinson to Rose Hill   *209*

30 The Lower Wabash Considered
    Beall Woods State Park and Environs   *217*

31 Caverns, Sinkholes, and Falling Water: The World According to Karst
    Dupo to Illinois Caverns State Natural Area   *223*

32 A Fault-Seeking Expedition
    Fountain Bluff to the Pine Hills Escarpment   *231*

33 Sandstone Gorges, Rock Shelters, and a Natural Bridge
    Bell Smith Springs   *239*

34 What Goes Down Must Come Up
    The Garden of the Gods and the Eagle Valley Syncline   *245*

35 The Geologic Wonders of an (Almost) Undiscovered Country
    Hicks Dome to Cave-in-Rock State Park   *253*

36 The Little River That Couldn't
    Hamletsburg to Cache River State Natural Area   *263*

37 The Meeting of the Waters
    Olmsted to Cairo   *271*

Glossary   *279*

Annotated Bibliography   *287*

Index   *295*

PREFACE

Illinois is a flat and boring state, pure and simple. It's a place to get through as quickly as possible. It is nothing but cornfields and crowded expressways. If you want to see beautiful scenery, go somewhere else. If you're interested in geology or the history of the earth, you'd better head to the Rockies, the Grand Canyon, or New England. Nothing important ever happened here, at least not until Abraham Lincoln decided to run for Congress.

The first paragraph is patent nonsense, supported only by willful ignorance and a systematic lack of curiosity. Yet every scientist and naturalist who has had the good fortune to see the amazing diversity of the Prairie State must endure this prevailing mindset at times. In fact, Illinois is a land of canyons and waterfalls, of broad, sky-filled expanses that contain a thousand doorways to vanished worlds. It is a land of the Pleistocene ice age, of giant Silurian reefs, of steaming tropical jungles, of soaring bluffs and deep rift systems, and of small streams in great valleys and huge rivers in narrow, rocky gorges. No vision of the West's mountains, no explanation of the East's plateaus, makes perfect sense without an understanding of this great land between. Scratch its surface anywhere and you will find more than great cities and fertile heartland. You will find a dramatic story stretching back more than a billion years.

This book is not intended to be a comprehensive guide to everything of geologic interest in Illinois. In your local park, on a short drive out into the country, or even in your own driveway, you can find fascinating geologic processes at work. Use this guide as a starting point. I will rest satisfied if the book whets your curiosity and causes you to reach beyond its scope.

ACKNOWLEDGMENTS

While compiling information for this book, I have learned to be more than a little ashamed of how long it took to fully appreciate my native state. As it turned out, no personal exploration has ever been happier or more fruitful. I can only hope that this book can transmit one ten-thousandth the sense of wonder—and one-thousandth the information—it has provided me. Given the exigencies of space and format, it contains only a fraction of what I have learned from people more skilled and perceptive than myself.

These people include past generations of geologists, paleontologists, and other scientists who played a major role in unlocking the secrets of Illinois's far-distant past. At least as important, though, are the people who have done much to help me find my way on this journey of personal discovery. First among them are the women and men of the Illinois State Geological Survey (ISGS), who encouraged my efforts and, in the midst of their busy schedules of field, research, and educational work, spent hours reviewing my text. Their number includes distinguished specialists in several fields who provide an impressive example of professionalism and commitment to public outreach. I encourage all my fellow citizens of Illinois to explore the excellent resource they have in the ISGS. Everyone who has an interest in the state's natural history (especially educators and their students) may attend the Survey's free public field trips and take advantage of its booklets, brochures, maps, and newsletters. Many of these are available in local libraries, at the Illinois State Library in Springfield, and at the ISGS public-information office at 615 East Peabody Drive, Champaign 61820.

For this book project, my primary contact at the ISGS was Myrna M. Killey, of the Quaternary Geology Section. As a geologist-educator with much experience in explaining both the thrill and intricacies of Illinois geology to the public, Myrna at once understood what can and cannot go in a nontechnical earth-science book, and proved to be an excellent editor. Her emendations and corrections to the text were delightfully perceptive and right on the mark. Further, she did a top-notch job in orchestrating the distribution and retrieval of appropriate text sections to and from other staff members. Throughout this long process, she maintained a sense of perspective and an appreciation for my point of view, which I found most heartening.

Among the other ISGS scientists who read my manuscript, I especially want to thank Ardith K. Hansel, a scientist with the Quaternary Geology Section, who is one of the foremost modern-day interpreters of the Pleistocene-Holocene history of northeastern Illinois and the Lake Michigan basin. Not only did Ardith provide me with an up-to-date Pleistocene-Holocene chronology; she also went out of her way to check other facts and to show my text to colleagues elsewhere.

In addition, I received thoughtful criticism, encouragement, and supporting literature or illustrations from the following ISGS geologists: Dennis R. Kolata, head of the Sedimentary and Crustal Processes Section; W. John Nelson and Janis D. Treworgy, both of the Sedimentary and Crustal Processes Section; Colin G. Treworgy, of the Coal Section; Joan E. Crockett, of the Oil and Gas Section; Donald G. Mikulic and John M. Masters, both of the Industrial Minerals and Mineral Economics Section; C. Pius Weibel, of the Quaternary Geology Section; and Wayne T. Frankie, head of the Educational Extension. All these scientists made special effort to help me understand the latest findings in their fields. If space limitations did not forbid, I would write an appreciative description of each of their contributions. It has been an honor to learn from them.

I also had the benefit of learning from scientists and naturalists in other organizations and institutions. My special thanks to W. Hilton Johnson, recently retired from the University of Illinois Department of Geology. Professor Johnson gave me a much clearer picture of Quaternary history of the Illinois River—a particularly dramatic and complicated subject. Furthermore, he acquainted me with Edwin Hajic's doctoral dissertation, which I found enlightening. In a different area, Eric Livingston, Chief Geologist of the Ozark-Mahoning Company, reviewed my essay on Hardin County. When I met with him, he gave a wealth of information, not only on his company's fluorspar mining, but also on other aspects of the complex geology of southern Illinois. And, I extend my gratitude to former Illinois State Museum colleagues: Librarian Orvetta Robinson, who located much important research material for me, and Geology and Paleobotany Curator Richard Leary. Richard provided me with much insight into the paleoecology and the fossil plants and animals of the Prairie State. He also steered me to publications and research materials that were hitherto unknown to me.

Many of the sites described in this book are state or county parks. To make sure that I had my facts straight, I contacted managers and naturalists at these locations whenever possible. I was encouraged by their help and positive response. These professionals include the following officials of the Illinois Department of Natural Resources at Beall Woods State Park: Site Interpreter Faye Frankland, Site Superintendent Jack Rhinehart, and Site Technician Joe Johnson. Separately, I was assisted by Heller Nature Center supervisor Steve Barg, as well as by other Park District of Highland Park personnel. And a special note of appreciation to the staff of the McHenry

County Conservation District, the agency that administers one of my favorite geologic sites, Glacial Park. Staff Ecologist Ed Collins read its essay, made important additions, and gave me a Cook's tour of the park and environs. Education Program Coordinators Katherine Beall-Ellinghausen and Linda Jaracz supplied other important information about the park.

In researching and writing this book, I have done my utmost to sift a coherent set of themes out of a mountain of information that describes many different subjects, and many different processes that took place over hundreds of millions of years. The information itself was published over the course of decades. I have had to decide how to selectively present older (and now often superseded) ideas in the context of newer ones, as a sort of commentary on how geologists refine or otherwise alter their theories of the world. I also had the heart-wrenching task of eliminating several important concepts, and superb geologic sites, from the final manuscript. As noted above, I have received a remarkable amount of help from experts in all these undertakings. The fact remains, however, that any attempt at interpretation or generalization of such a complex subject is bound to disagree here and there with other points of view. Accordingly, my opinions, and any remaining factual mistakes, are solely my own.

Era	Period	Significant Developments	Millions of years ago
Cenozoic	Quaternary	The Pleistocene and Holocene Ice Age: at maximum extent, about nine-tenths of Illinois covered by ice.	
Cenozoic	Tertiary	The Paleocene, Eocene, Oligocene, Miocene, and Pliocene epochs. During at least the first two epochs, the continued deposition of sediments in southernmost Illinois.	1.6
Mesozoic	Cretaceous	Beginning of Mississippi Embayment and renewed marine deposition in southernmost Illinois. Rise of flowering plants. Disappearance of dinosaurs at end of period.	66
Mesozoic	Jurassic	Illinois still dry land: no surviving rocks or fossils. Heyday of the dinosaurs.	144
Mesozoic	Triassic	Pangaea begins to break up by the end of this period. Illinois probably remains above sea level: no surviving rocks or fossils.	208
Paleozoic	Permian	No surviving rocks or fossils in Illinois: the terrain is above sea level. Pangaea in place.	245
Paleozoic	Pennsylvanian	The formation of the supercontinent Pangaea. The time of great tropical swamp forests. Extensive fossil record in Illinois.	286
Paleozoic	Mississippian	A period of shifting coastlines and deposition patterns: limestones and the Warsaw shale form.	320
Paleozoic	Devonian	The rise of trees and extensive land vegetation. Some marine deposition in Illinois.	360
Paleozoic	Silurian	The formation of coral reefs in Illinois. The Midwest is situated south of the equator.	408
Paleozoic	Ordovician	A shallow sea covers Illinois. Deposition of the widespread St. Peter sandstone and carbonate rock units.	438
Paleozoic	Cambrian	Rifting in southernmost Illinois ends. Illinois Basin begins to form as crust gradually subsides. A shallow sea covers part of Illinois for part of the period.	505
	Precambrian time (almost nine-tenths of earth history)	Continental rifting in southernmost Illinois begins. Evidence of organisms with advanced cell structure. Igneous rocks now constituting Illinois's Precambrian basement begin to form. Illinois terrain is bare and hilly. Evidence of primitive organisms. Formation of earth's crust.	570 / 1,400 / 3,600 / 4,600

Geologic time scale noting significant developments in Illinois.

Illinois Back When
AN OVERVIEW

What liberates is the knowledge of who we were, what we became;
where we were, whereinto we have been thrown; whereto we speed,
wherefrom we are redeemed; what birth is, and what rebirth.

> —A formula of the early Christian
> Valentinian Gnostics, ca. second century A.D.

In its broadest sense, geology is a field of inquiry that extends the human passion for understanding history far back beyond our own beginnings. To fully appreciate the sites described in the following essays, it is helpful—indeed, essential—to have some grasp of the major benchmarks of the vast span of geologic time. It is one thing to memorize the names and dates of the eras, periods, and epochs; it is quite another, given our fleeting existence as individuals, to really comprehend a quantity as great as a million or a billion years. It requires a profound shift in thinking, which for some people can be difficult. We live in a world where the entire record of human existence is the slimmest footnote of the story.

No place on the planet can boast a complete, well-preserved record of earth history from the planet's beginning to the present day. In Illinois, for example, certain periods and epochs are abundantly represented; others are missing entirely. Fortunately, there are instances when geologists can draw conclusions not only from rocks and fossils but also from their absence. They can also refer to evidence from neighboring areas, and even from far-distant locations. Consequently, it is possible to paint a general picture of Illinois through the ages.

Precambrian Time (4.6 billion–570 million years ago)

The Precambrian, the first span of geologic time, preceded the wide-spread appearance of complex, multicellular organisms in the fossil record. It is difficult to piece together many of the details of that time—a sobering fact, given that it makes up nine-tenths of the entire geologic time scale. But as earth scientists and physicists have refined their theories, discarded flawed assumptions, and improved their dating techniques, they have continued to revise the postulated date for the origin of the earth. Most experts today believe that our planet's crust had formed by about 4.6 billion years ago.

Because the earth as a whole is such a dynamic environment, and because it often recycles its raw materials, it is impossible to find a rock outcrop anywhere that closely approaches that almost incomprehensibly distant date. Luckily, some of the world's oldest unaltered sedimentary rocks, located in Australia and Africa, reveal an amazing fact: rudimentary forms of life were already inhabiting the ocean by 3.6 billion years ago. However, it appears that evolution proceeded very slowly at first, because there is no proof of organisms with advanced cell structure before the 1.4-billion-year mark. The record also suggests that there were no multicellular creatures in any abundance until the end of the Precambrian.

The most ancient Precambrian rocks discovered in North America are the heavily metamorphosed Isua series of western Greenland. Isotope dating places them at about 3.75 billion years. In Canada, Minnesota, and Wisconsin, Precambrian formations are exposed extensively at the surface. The oldest of these are located mainly toward the center of what geologists call the Canadian Shield—the ancient nucleus of our continent. Regrettably, no Precambrian rocks outcrop within the boundaries of the Prairie State. To see the closest exposure of that kind, you must travel north of the state line about 60 miles, to the Baraboo area of Wisconsin. Still, all of Illinois is underlain—though often at great depth—by what is called the Precambrian basement. This continuous surface of crystalline rocks 1.5 to 1.0 billion years old is composed primarily of granite, one of the commonest types of igneous rock found in continental crust.

In northwestern Illinois, the Precambrian basement lies a little more than 2,000 feet below the surface; in Chicagoland, about 4,000 feet down; and in the southernmost part of the state, as deep as 17,000 feet. However, this complex did once see the light of day. Near the tail end of the Precambrian, it was exposed to the forces of erosion. Imagine standing on this igneous version of Illinois: all around you stretches a bare, hilly landscape. There are rivers and smaller streams; but there is no vegetation or sign of animal life. Even the atmosphere is inhospitable, because persistent free oxygen, the gift of the abundant photosynthetic organisms yet to come, has not accumulated in any quantity. In the southernmost section of the state, you would encounter rift valleys similar to those of East Africa today. This zone of crustal weakness, which geologists have named the Rough Creek graben-Reelfoot rift, probably represented one relatively small episode in the breakup of a supercontinent that existed in the late Precambrian.

The Paleozoic Era (570 million–245 million years ago)

Illinois's complete lack of Precambrian outcrops is more than compensated by its rich store of Paleozoic formations. The largest and most important structural feature in the state, the Illinois Basin, is a product of the Paleozoic. Apparently, it continued to subside sporadically throughout most of the era, as it received layer after layer of new sediments. The basin, centered in the southeastern counties, forms a huge oval accumulation of down-ward- and inward-dipping rock strata. In its thickest portion, in northern Pope County, it is more than 3 miles deep.

The Paleozoic is made up of seven main subdivisions, called periods. The first, the Cambrian period, lasted from 570 to 505 million years ago. Marine sedimentary rocks of Cambrian age outcrop on private land in Ogle County, but the exposures are not particularly easy for the layperson to find. Happily, the record of the next period, the Ordovician (505–438 million years ago), is much better represented. During much of this time, a shallow sea covered the state. One of the Midwest's most extensive and most eco-nomically important sedimentary formations, the St. Peter sandstone, was deposited in a huge apron around the Ordovician coastline; later, limestones and dolomites formed in deep-water conditions. Ordovician outcrops are now especially common in the state's northwestern and north-central counties, and along several stretches of the Mississippi River bluff in the

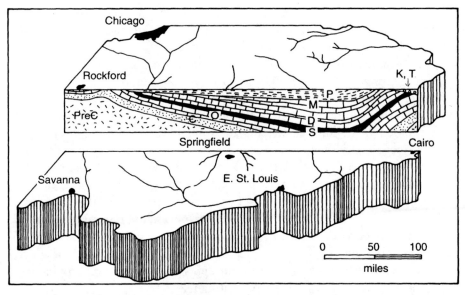

The downward-arching subsurface feature known as the Illinois Basin, a vast accumulation of hundreds of millions of years of sedimentary deposits. The letters refer to the geologic ages of the rocks: Pre-Є=Precambrian, Є=Cambrian, O=Ordovician, S=Silurian, D=Devonian, M=Mississippian, P=Pennsylvanian, K=Cretaceous, and T=Tertiary. —Illinois State Geological Survey

3

Fabricated stepping-stones lead visitors across a canyon creek at Matthiessen State Park. The park's spectacular dells are cut into Ordovician St. Peter sandstone.

southern half of the state. Strata laid down in this period have proven to be an important source of lead and zinc ores in the Galena area, and of lime, crushed stone, and crude oil in other locales.

The Silurian, a relatively short period that lasted from 438 to 408 million years ago, provided modern Illinois with resistant dolomite, a magnesium-bearing carbonate rock that forms the tops of the state's highest elevations in Jo Daviess County, as well as the spectacular Mississippi Palisades bluff near Savanna. The Silurian was also the time when marine organisms built many reefs in the shallow saltwater sea. Some of them, most notably the Thornton Quarry reef on Chicago's south side, were huge structures that have given generations of paleontologists a detailed view of the ancient marine environment. The following Devonian period (408–360 million years ago) is not so well represented as the Silurian period in Illinois, but its cherts and limestones form part of the Mississippi Valley cliff line, particularly at the beautiful Pine Hills Escarpment and the Devil's Backbone, both of which are located in Jackson County. In some spots, Devonian rocks provide tripoli, a commercially valuable silica substance.

The next two Paleozoic periods, the Mississippian (360–320 million years ago) and the Pennsylvanian (320–286 million years ago), are sometimes grouped into a single period, named the Carboniferous. (In this book, I conform to the prevailing American usage and keep them distinct.) In the Mississippian, Illinois was once again mainly a seawater environment, but

Cleft phlox plants blooming in early spring amid the steeply dipping beds of Devonian limestone at the Devil's Backbone in Jackson County.

the coastline to the north and northeast shifted a great distance up and down. The rising and falling sea level was apparently triggered by changes in Illinois Basin deposition and subsidence, and perhaps by the slow movement of continents into former oceanic areas. Our state's large assemblage of Mississippian strata comprises a variety of sedimentary types, but chief among them is limestone. This rock is quarried extensively for road and building construction, and in some locations it is also a significant reservoir of crude oil. In Illinois, Mississippian rocks form part of the uplands of the far-southern counties and are also extensively exposed along the bluffs of the Mississippi and lower Illinois Rivers.

The geologic and economic significance of these earlier Paleozoic rock strata is impressive, but Illinois is most famous for its Pennsylvanian formations. Not only do these rocks contain the state's number-one mineral resource, coal; they also preserve a superb fossil record that has provided us with an excellent look at the plants and animals of a profoundly different world. This period was a time apart, when conditions in any one locale changed with a frequency unmatched in other times. As a result, many individual Pennsylvanian beds are relatively thin: sandstone quickly gives way to shale; shale soon gives way to limestone; and so on. This rapid succession of the rock types falls into a series of repeating patterns, which Illinois geologists refer to as cyclothems.

Masses of Pennsylvanian sandstone rise above a forested, hilly landscape at the Garden of the Gods, in Shawnee National Forest.

In Pennsylvanian time, our continent was positioned astride the equator. It was also in the process of colliding with Africa and Europe—an event that was part of the even larger coming-together of the supercontinent Pangaea (pronounced pan-*jee*-ah). Many geologists now believe that the stresses produced by this upheaval left their mark this far into the American heartland. If so, the fault systems and zones of crustal deformation found in various parts of Illinois bear witness to the great changes under way 1,000 miles to the east. The Midwestern landscape of the Pennsylvanian period was low-lying and mantled in lush tropical growth—seed ferns, true ferns, and giant relatives of modern horsetails and club mosses. This swampy world was also inhabited by primitive forms of insects, amphibians, and early reptiles. As plant parts and even whole forests succumbed to the latest incursion of the sea, they fell where they had stood, and formed thick layers of peat. In the course of the ages, this peat was transformed by compression and heat into the state's extensive deposits of bituminous coal.

The final Paleozoic period, the Permian (286–245 million years ago), was the only one that left no lasting trace of its passing in this state. If any Permian beds were deposited, they have since been removed by the relentless forces of erosion. Our continent was firmly locked into the main body of Pangaea, which extended virtually from one pole to the other.

The Mesozoic Era (245–66 million years ago)

"Were there ever any dinosaurs here?" Most Prairie State geologists have been asked this question at one time or another. Unfortunately, it can't be answered with certainty, because the Mesozoic, otherwise known as the Age of Reptiles, is not well represented in Illinois. Rocks and other sedimentary deposits from the first two Mesozoic periods, the Triassic (245–208 million years ago) and the Jurassic (208–144 million years ago), are wholly lacking. In all likelihood, the entire expanse of the state was above sea level; so it's difficult to imagine that dinosaurs didn't inhabit it. Intriguingly, there are significant deposits of sediments from the third and final Mesozoic period, the Cretaceous (144–66 million years ago), in west-central and southernmost Illinois. But while they contain some excellent plant fossils, they have not yet yielded even the tiniest dinosaur bone. (Tantalizingly, though, dinosaur bones have been discovered in the same sediments in Tennessee and Missouri.)

The Cretaceous sediments—they never were lithified, or transformed into rock—were laid down in conditions that were well on their way to becoming what we consider normal for Illinois and the continent in general. It was still a warm, subtropical environment, but North America had

Bald cypress trees thrive in the Coastal Plain environment of Horseshoe Lake, located north of Cairo in southern Illinois. The lake lies in an abandoned meander of the Mississippi River.

7

broken free from the more eastern landmasses, and the new Atlantic Ocean was widening. In the South and Midwest, however, there was one particularly exotic feature: the sea extended up a narrow corridor to the southern tip of Illinois. This incursion of salt water was caused by the subsidence of a new structure, the troughlike Mississippi Embayment, situated where the lower Mississippi River now flows.

The Cenozoic Era (66 million years ago to the present)

The Cenozoic is an awkward part of the geologic time scale, because it is split into two periods of vastly different sizes: the Tertiary (66–1.6 million years ago) and the Quaternary (1.6 million years ago to the present). The Tertiary period began with an arm of the sea still lapping over the state's southern tip. Marine sediments were deposited in the Cairo area in the Paleocene and Eocene epochs, the two earliest subdivisions of the Tertiary. One of these formations yields a special, highly absorbent clay that in modern times is processed and marketed as cat box filler. Much later—probably toward the end of the Tertiary and long after the sea had withdrawn southward—a river system deposited layers of gravel in at least several different areas of the state. That gravel is best seen in outcrops in Illinois's far-southern counties.

In geologic terms, the Quaternary began just a blink of an eye ago, but what an eventful period it has been already. Despite its short duration, it has been such an important episode in earth history that it has been given two subdivisions: the Pleistocene epoch (1.6 million–10,000 years ago) and the Holocene epoch (10,000 years ago to present). Originally, scientists thought that the Pleistocene was the Ice Age. But evidence now suggests that the Holocene is a completely artificial unit of time, nothing more than one of the warm, interglacial spells in a larger, continuing event. This new interpretation is partially based on a better understanding of the typical 10-million-year length of earth's ice ages. (There appear to have been at least three others: one in the middle of the Precambrian, another near the end of the Precambrian, and one in the Permian period. We have no proof that these earlier ice advances reached ancient Illinois.)

Whatever the validity of the Pleistocene-Holocene distinction, most geologists still honor it. Traditionally, the Pleistocene was further broken up into stages, including four—the Nebraskan, the Kansan, the Illinoian, and the Wisconsinan—that supposedly corresponded to the major glaciations, or large-scale advances of the ice sheet. We now know, however, that up to twenty such glaciations have taken place in the most recent ice age. In an effort to keep a reasonable nomenclatural balance between customary usage and the new discoveries, Midwestern scientists have jettisoned the Nebraskan- and Kansan-stage names and replaced them with the blanket term "Pre-Illinoian." The Pre-Illinoian stage (1.6 million–300,000 years ago) comprises the bulk of the Pleistocene. It was followed by the last two glaciations, the Illinoian (300,000–125,000 years ago) and the Wisconsinan (75,000–10,000 years ago).

The record of Pre-Illinoian ice advances into the Prairie State is not abundantly clear. Later glaciations obliterated some of it, and there has been a long time for the oldest glacial drift to be removed by wind and running water. Nevertheless, it is clear that early glaciers invaded the state from both the west and the east in Pre-Illinoian time. Probably more than one advance reached the central counties.

The legacy of the Illinoian stage was more extensive and more clearcut. During this time, the continental ice sheet reached its southernmost extension anywhere in the northern hemisphere. That southernmost point

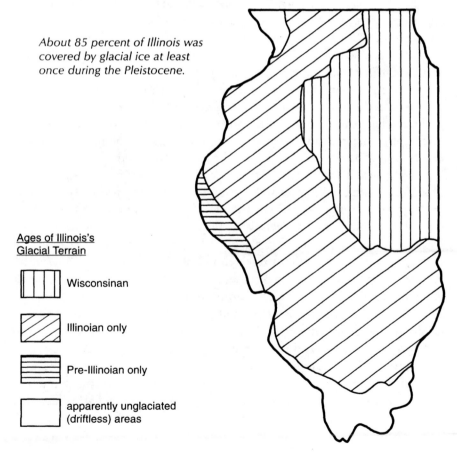

About 85 percent of Illinois was covered by glacial ice at least once during the Pleistocene.

Ages of Illinois's
Glacial Terrain

Wisconsinan

Illinoian only

Pre-Illinoian only

apparently unglaciated
(driftless) areas

9

Chicagoland is one of several urban areas in the state that is experiencing an increase of new residential neighborhoods. This clonescape addition is on the flat floor of Lily Cache Slough, on Chicago's far southwest side. At one time, this area was a major sluiceway for the draining waters of ancient Lake Chicago.

was in Illinois, a few miles north of the Shawnee Hills. Largely because of the far-ranging Illinoian ice lobes, about 85 percent of the Prairie State experienced the full brunt of the glaciers. The Wisconsinan glaciation that concluded the Pleistocene did not make it as far, but it left the best glacial landforms and sediment deposits of all, in the northern half of the state. The complex pattern of end moraines and till plains—not to mention modern Lake Michigan—are all leavings, so to speak, of the Wisconsinan stage.

Debate continues about what triggered the Pleistocene ice age. The periodic changes of the so-called Milankovitch cycle—a repeating pattern of climatic changes caused by astronomical factors involving the shape of earth's orbit, the tilt of its axis, and the precession of the equinoxes—apparently played a major role in triggering the buildup of far-spreading ice caps. Perhaps the buildup was also encouraged by a series of sudden shifts in the course of the North Atlantic's Gulf Stream. There are many other suggested causes as well: reversals in the earth's magnetic field, increased volcanic activity, the arrival of galactic dust clouds, and so on. Whatever the ultimate causes were, modern research reveals that the ebb and flow of the ice sheets was not the result of gradual rises and drops in average annual temperature, but of surprisingly abrupt and dramatic climatic changes. It also appears that the ice sheets started growing not in times of especially harsh winters, as one might suppose, but in spells when the summers were unusually cool. Low maximum temperatures, not low minimum temperatures, provided the conditions necessary for the onset of one of the most unusual happenings in all earth history. As strange as it seems, winters in Illinois during the glacial advances were probably somewhat milder than the ones we experience in modern times.

Tips for Exploring the Sites
Described in This Book

*To everything there is a season, and a time to every
purpose under heaven.*
                                        —Ecclesiastes 3:1

— Late fall, after leaf drop, and early spring, before bud break, are the best
times to explore Illinois's geology. You will get a much better view of
the sites at a time when they are not cloaked by heavy vegetation. At
these times of year, however, back roads (particularly unpaved ones)
may be muddy or badly eroded. Be alert, so you won't get stuck far
from the nearest town.

— Collect specimens only where permitted. Specimen collecting is for-
bidden by law in most federal, state, and county preserves. In a few
places, such as Geode Park in Warsaw, Illinois, collecting is permitted.
But in general, do not remove rock or fossil samples from public parkland
without the express permission of the park's administering agency.

— Be extremely careful. Do not use a rock hammer unless you are wearing
safety goggles. Also, wherever you find yourself face-to-face with a rock
outcrop, beware of rock slides and rocks falling from above.

— A few of the sites in this book are situated on potentially dangerous
clifftops, boardwalks, or riverbanks where quicksand conditions and rapid
flooding have been observed. Especially in southern Illinois, some sites
are home to venomous snakes. Use common sense and look where
you're stepping.

— In the fall and spring, rubberized ooze shoes (preferably high-topped)
are indispensable.

— If you've never been on a geologic expedition before, consider attend-
ing at least one of the several excellent (and free) earth-science tours
presented each year by the Illinois State Geological Survey. You will
learn a great deal and visit fascinating sites normally closed to the public,
and you will meet some of the experts who are at the forefront of their
fields. For tour schedules and more information, contact the Survey's
Educational Extension at 615 East Peabody Drive, Champaign, IL 61820.

11

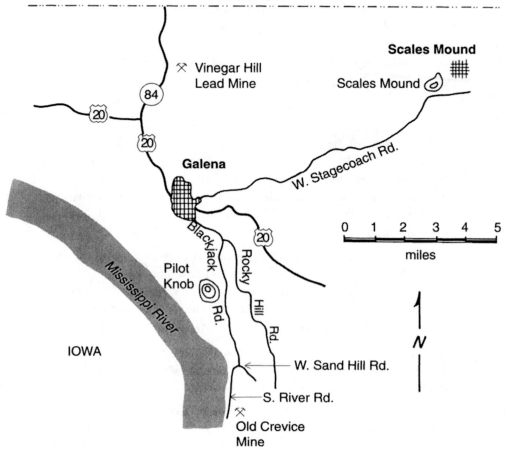

Galena and environs.

— 1 —
Lead Mines and Lovely Hills
GALENA AND ENVIRONS
Jo Daviess County

There is something fascinating about science. One gets such wholesale returns of conjecture out of such a trifling investment of fact.
—Mark Twain, Life on the Mississippi

Geology is a fascinating and often frustrating field of endeavor. More often than not, its practitioners have to deal with Twain's trifling investment of fact—in the form of scattered rock outcrops, incomplete fossil evidence, or patterns in the landscape that can mean any one of a number of things. But because it is their job to create a convincing model of the past, geologists must sometimes produce wholesale returns of conjecture, in the understanding that they may never really know if they are right. The earth is not a well-designed laboratory; unlike scientists in many other disciplines, geologists rarely have the luxury of running carefully controlled experiments that thoroughly prove or disprove their theories. Years of accumulated experience and reflection, though, coupled with the use of modern investigative tools and techniques, can give geologists almost uncanny interpretive skills and a mastery of their subject that the rest of us can only marvel at.

In this essay, the first of a pair devoted to Illinois's incomparable Jo Daviess County, you'll see how geologists have worked with clues from the past to deduce the origins of this lovely, hilly landscape and of the glittering metal ore that once drew so many people to it. In modern times, most people visit Illinois's northwesternmost county for one of two reasons: to enjoy the area's fetching Wisconsinesque scenery or to browse for antiques. Visitors 150 years ago had a different agenda. Unlike today's day-trippers from Windy City suburbs, they came not from due east, but from points south, traveling up early America's best approximation of an interstate—the Mississippi River.

The treasure they sought was not vintage farmhouse furniture or a few hours' escape from urban sprawl. Indeed, in 1827, when the city of Galena was officially chartered, it was a more populous and bustling place than the Chicago of that day. The treasure the new arrivals sought was something much more in demand in early America: the ore containing lead, a metal once used in everything from printer's type and house paint to water pipes and musket balls. The early-nineteenth-century surge to exploit the rich mineral resources of Jo Daviess County resulted in one of the nation's first "rushes," beginning more than two decades before the discovery of gold in California.

To the geologist, however, the history of local ore-mining is one of many fascinating aspects of this beautiful corner of Illinois. The area is part of a region geographers call the Wisconsin Driftless Section—a unique part of the Upper Midwest that apparently never suffered the direct onslaught of the Pleistocene glaciers. (I say apparently, because Illinois State Geological Survey staff members have reported the presence of some erratics, or glacially transported rocks, near Galena and Hanover. These exotic rocks, together with some other glacial sediments found atop the Mississippi bluff near East Dubuque, imply that an ice-sheet intrusion took place here early in the Pleistocene. But if any glacier reached the Driftless Section, it left no other surviving evidence.)

The region's penchant for avoiding all or most of the ice sheets was possibly due to its location south of a high area in Wisconsin. While we may be tempted to think that Jo Daviess County is a well-preserved "fossil landscape," giving us a detailed picture of how Illinois looked before the Pleistocene, caution is indicated. As we will see in essay 2, the region was profoundly affected by the Ice Age, even if it did not directly feel the weight of the ice. Pleistocene changes in erosional patterns and river systems left their mark even here.

One of the best places to study the geologic aspects of this terrain is also one of the best spots to enjoy its refreshingly nonlinear scenery. A few miles northeast of Galena, on West Stagecoach Road heading toward Scales Mound, is a fine stretch of the Wisconsin Driftless Section. (As its name suggests, most of the section is in Wisconsin; the term *driftless* refers to the fact that almost no drift, or glacier-borne debris, has been found here.) For decades, scientists have been debating whether this region contains remnants of one or more peneplains. A peneplain is a low, gently rolling landscape that marks the final stage in a cycle of erosion. In this condition, local streams have finished their slow but determined work of removing the high ground; they have virtually reached the magic limit, known as base level, below which they may not cut.

It is a brutal fact of life, though, and particularly of geology, that nothing is that elegantly simple. Landscapes do not inexorably start with high relief, get beveled flat, and stay there for the rest of eternity. Often a tug-of-war is at work between the forces of erosion and uplift. For instance, it's possible for an area to be eroded down to a fair approximation of a peneplain only to become relatively high ground again—either because it has been physically lifted upward by forces in the earth's crust, or because factors elsewhere have altered the base level. In any event, the streams renew their downcutting. In time, only a few disconnected fragments of once-continuous peneplain may remain. Geomorphologists (the specialists who study the evolution of landforms) snoop about continually in search of these old dismembered peneplains; and here, sure enough, they think they've found one, and probably even two.

The older of these two proposed peneplains, the Dodgeville, is the more controversial. Its proponents say that it came into being in the Pliocene, the epoch directly preceding the Pleistocene ice age. They regard the tops of the region's highest hills, which gently tilt southwestward from about 1,200 to 1,000 feet above sea level, as all that remains of the Dodgeville surface in Illinois. You can spot this putative peneplain along West Stagecoach Road as you approach the hill called Scales Mound (it stands just southwest of the town by the same name). Use your mind's eye to connect its top with the summits of the other high hills in the distance. As you drive up and over the southern slope of Scales Mound, you'll see its cap of resistant Silurian-period dolomite. This gray-toned sedimentary rock—especially tough stuff, thanks to its densely packed crystals—is the youngest formation in the area, being a mere 430 million years old or so. Still, it is roughly one hundred times as old as the peneplain it now helps to define.

The summit of Scales Mound. Note the cap of thin-bedded, cherty Silurian dolomite.

The view to the northwest from Blackjack Road. The cattle stand on the Lancaster peneplain surface underlain by Ordovician Galena group dolomites. Across the Mississippi in Iowa stands a higher erosion surface, the Dodgeville peneplain.

After persuading yourself that the Dodgeville peneplain really exists, consider the case of geologist A. C. Trowbridge. In 1921 he first proposed the existence of the Dodgeville, in a preeminent study of the Driftless Section. Thirty-four years later, having persuaded most of his colleagues in the meantime, he changed his mind. (Unrepentant advocates of the Dodgeville can take heart from Trowbridge's contemporary, Leland Horberg. One of Illinois's most respected geologists, Horberg remained firm in his belief that the Dodgeville was a true entity.)

The second peneplain, the Lancaster, was created after another episode of uplift or base-level change. It came into being after the Dodgeville, but it too is probably a product of the Pliocene. In this area it marks the upper surface of the Ordovician-period rocks of the Galena group. As you drive northeast of Galena and watch the level of the land in the distance, you should be able to recognize the Lancaster as a less than perfectly level surface, about 200 feet below the Dodgeville. It is also revealed on the lower hilltops that Route 20 passes over on its way from Elizabeth to Woodbine. The Lancaster peneplain is believed to be part of a much vaster erosional surface that can still be found, under different names, in the Ozarks, in Tennessee, and in Kentucky. Farther south in our state, it manifests itself as the Calhoun peneplain (see essay 24). Just to the east, in Stephenson County, it actually forms the general upland surface, though it is thinly mantled by glacial drift.

Once you've done your best at peneplain-hunting, head back toward Galena to consider the somewhat less abstruse subject of this region's mining history. A stroll through the town's central area, even on the busiest of weekends, is a delight. Not only is it a fascinating excursion into the life and times of that hometown-failure-turned-national-hero, Ulysses S. Grant; it is also a monument to the wealth and refinement enjoyed by at least some of its other residents in the nineteenth century. This prosperity was the direct result of the lead and zinc deposits concentrated in Jo Daviess County, mainly in a zone about 15 miles long by 10 miles wide.

Two years ago, when I was visiting Galena, I stopped in a knickknack shop and purchased a small, cube-shaped lump, as shiny as freshly polished silver but much heavier. The souvenir sits on my computer monitor as I write this: it is a specimen of lead sulfide, otherwise known as galena, the city's namesake. It is this dense and eye-catching mineral that was so eagerly sought by the region's early miners—not for its aesthetic appeal, but because it is the source of lead.

By no later than the 1500s, the American Indians living on the Iowa side of the river were mining lead deposits there. This activity came to the attention of French trader Nicolas Perrot in 1682; but while there was some on-again, off-again mining in the region in the century and a half that followed, it was not until the early 1820s that the "lead rush" really began. From that point until roughly the end of the Civil War, Galena and its environs were the nation's lead-production center. In 1845, halfway through the boom, this mining district supplied nine-tenths of America's lead needs.

At first, the miners concentrated on the lead ore and did not devote their efforts to one of its associated minerals, sphalerite, a chief ore of the metal zinc. Many of the lead deposits lay closer to the surface and were easier to extract, and an economically viable method of processing sphalerite was not developed until 1858. At that point, lead was widely marketable and zinc was not. Zinc production eventually outstripped lead, at least until 1973, when Jo Daviess County's last commercial mining operation closed down for good.

The first question a geologically inquisitive person is likely to ask about Galena is, Why did the lead and zinc deposits occur here, and not elsewhere in Illinois? The preponderance of dolomite in the area's bedrock probably had something to do with it, even if it was only one of several determining factors. Some of the world's other giant lead-ore deposits, including those in southeastern Missouri and Poland's Upper Silesia district, are in dolomite belts. But many other dolomite belts in the Prairie State—Chicagoland, for instance—apparently contain no ore at all.

Geologists used to theorize that the lead and zinc had been in the dolomite all along. They reasoned that groundwater moving through rock strata gradually dissolved the metals, then redeposited them in more concentrated form, in cracks and crevices in the strata. Nowadays, most geologists favor the theory that the ores came from hydrothermal (warm-water) solutions produced elsewhere at great depth. According to this model, these solutions migrated a long distance through permeable rocks—in this case, perhaps the Cambrian sandstones that underlie the Ordovician strata—and then moved upward, through hard-to-detect faults. In the process, they lost some of their heat and in the end reached a zone where the temperature, pressure, and chemical conditions encouraged both precipitation of the ore and replacement of some of the host rock. What remains unclear is why the ore is restricted to one relatively narrow vertical section of all the region's Ordovician dolomite.

The ore deposits manifest themselves in two different forms: the more easily found and exploited crevice type, and the more deeply seated flat-and-pitch type. Lead ore is most commonly found in crevice deposits, which are located in widened joints in the upper part of the large Ordovician-age rock assemblage known as the Galena group. Zinc ore is concentrated in the flat-and-pitch deposits, which are mainly in the lower part of the Galena group, and in the even lower Platteville group, also of Ordovician age. Most often, flat-and-pitch deposits lie between the bedding planes of the strata, or slice through them at an angle greater than 45 degrees.

Most of the crevice deposits were mined out long ago. Periodically geophysicists and mining engineers try to locate new flat-and-pitch deposits with hi-tech prospecting equipment. Even if you lack their expertise and fancy gear, you can easily locate two old mines within a few minutes' drive from downtown Galena. The first is the privately owned Vinegar Hill Mine and Museum, located off Route 84 north of town. Signs will guide you to it, but note the facility is closed to the public in winter months.

The second site is a more primitive affair: an old crevice mine in the bluff fronting the Mississippi River. To reach it, take Blackjack Road south from Galena. (The first time I took this lovely, serpentine road, the botanical side of my brain wondered whether it was named after a northern population of blackjack oaks nearby. But the term *blackjack* was miners' slang for sphalerite—a more plausible explanation here.) After about 5 miles of winding every which way, turn right onto West Sand Hill Road. At the base of the bluff, turn downstream on South River Road. At 0.7 mile, on your left, you'll spot the exposed vertical shaft of a crevice mine. It is propped by a horizontal timber near its base. This bluff was a heavily worked mining area. Take a good look at it from the roadside, but do not enter this or any other mine shaft. They are dangerous.

When you return up Blackjack Road toward Galena, you can briefly revisit the first theme of this essay and do one more bit of peneplain-spotting. Across the river, to the northwest, the escarpment of Silurian rocks in Iowa marks the high surface of the Dodgeville peneplain; closer at hand, you'll travel over the Lancaster peneplain wherever the road takes the high ground atop the Galena group dolomites.

A nineteenth-century crevice mine in the Galena group dolomites along the Mississippi River bluff, south of Galena. A tree has managed to put down its roots into the shaft. The mine is unsafe to enter.

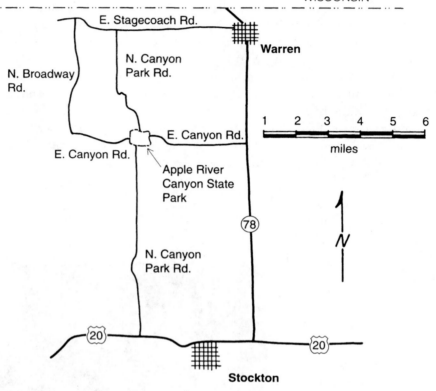

The easiest way to reach Apple River Canyon State Park is via the county's main east-west thoroughfare, U.S. 20. At the intersection about 2 miles west of Stockton, take North Canyon Park Road heading north.

— 2 —

A Diversionary Tale
APPLE RIVER CANYON STATE PARK
Jo Daviess County

The secret of the stars,—gravitation.
The secret of the earth,—layers of rock.
—Edgar Lee Masters,
Spoon River Anthology

One of the great paintings of the Italian Renaissance, Raphael's *School of Athens,* is a perfect metaphor for those scientists who seek the origins of our world. Two philosophers rapt in conversation are walking out of a gleaming temple: on one side, Plato, who, untouched by earthly concerns, has his hand pointed upward toward the heavens; on the other, Aristotle, the naturalist, gesturing outward and down to encompass the earth. Astronomers would claim this version of Plato for their role model, for in seeking the beginnings of the universe in its dance of gravitation they look upward to the ancient light of fast-receding galaxies. Geologists, also infatuated with beginnings, seek them as Aristotle would, by searching for the secrets that lie about and below, in our own planet's layers of rocks.

The sedimentary strata of Jo Daviess County have commanded the attention of generations of miners. But they are intriguing for another reason. They reveal vivid glimpses of the distant past. At Apple River Canyon State Park, one of Illinois's most secluded and ravishing natural settings, the record of the rocks is especially easy to scrutinize.

As you approach the park from the south on North Canyon Park Road, you'll pass through a portion of the Driftless Section, which stands only a couple of miles west of where the Pleistocene ice sheet of the widespread Illinoian glaciation had its farthest extension. After entering the park as you descend to the foot of the pretty Apple River, you pass through the silent millions of years recorded in layer after layer of Ordovician rocks. Take a good look at the canyon face north of the parking lot and the road junction, near the base of Primrose Cliff Nature Trail. What is the predominant rock type here? Is it soft and yielding, like shale, or does it more closely resemble the carbonate sedimentary rocks, limestone or dolomite? The vertical cliffs, a solid erosion surface, strongly suggest that this isn't shale. It is dolomite

21

The Ordovician Dunleith formation at the base of Primrose Cliff.
The dolomite strata look like they were laid down by a meticulous
stonemason. The center beds are less resistant to erosion than the
upper and lower beds.

of the Dunleith formation, a part of the Ordovician Galena group. This
rock, nicknamed "The Drab" by nineteenth-century miners, formed far
offshore approximately 450 million years ago, when Illinois was covered by
a shallow, saltwater sea. (For more on the origin of dolomite, and on how
it differs from the look-alike limestone, see essay 3.)

The Galena group is one of the main zones of lead and zinc ore deposits; and not surprisingly, there are quite a few old mine workings in this
area. Here, however, the cherty and brownish gray dolomite holds another
sort of treasure: the fossil named *Receptaculites*. Over the years, paleontologists have described it as a sponge or some other kind of marine invertebrate. It is probably a structure that was produced by algae. Its common
nickname, "sunflower coral," is because it has the shape and spiral pattern
of a sunflower's floral disk, but paleontologists are certain it was neither a
flowering plant nor a coral-building animal. Whatever the identification
problems, this distinctive organism is an excellent index fossil, and as such
it is used to date and identify rocks at different outcrops over a large area.

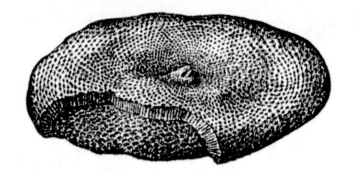

Receptaculites is found at three distinct and consistent levels in the Galena group. The specimens here belong to the middle zone.

Once you've scrutinized the Dunleith dolomites, stroll back across the bridge and walk down to water's edge in the picnic area, just west of the parking lot. Here, you are at the junction of the Apple River's main branch and its South Fork. The main branch comes in from the park's west side, combines with the South Fork, and then takes a sharp dogleg through the narrow canyon to the southwest. If you could hover high in the air over this site, you might notice several strange things about it. Upriver from the park, the main branch of the Apple acts like a normal river: its valley winds back and forth and widens as it descends, as do the region's other stream courses. But on reaching the dogleg and the canyon, the Apple suddenly

View from the Primrose Cliff Trail observation platform. This is the final stretch of the Apple River's older, winding, southeast-trending valley. Just past the bridge at upper right, the river makes a sharp dogleg to the right, through a steep-walled canyon.

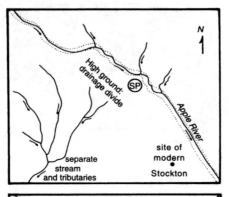

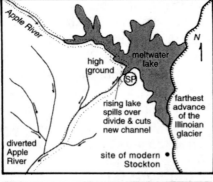

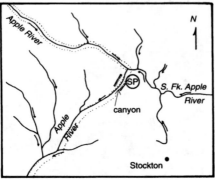

The Apple River diversion.
SP marks the site of Apple
River Canyon State Park.
—Illinois State Geological Survey

(Top) The Apple River's
northwest-to-southeast valley
before the Illinoian glaciation.

(Middle) A meltwater lake
formed at the margin of the
Illinoian glacier and caused
the Apple River to cut a
southwestward channel.

(Bottom) The Apple River's
modern course.

stops winding and follows an unusually steep, straight course for several miles before it resumes its old meandering ways. Further, there's something odd about the South Fork, too. Its valley narrows in the downstream direction; stranger yet, the tributaries to the east tend to enter it pointing upstream. To a geomorphologist, these telltale facts practically leap off the topographic map. They are classic indicators that a river, or at least a section of it, has reversed course at some point in the geologic past. This park is a famous example of the stream-reversal process.

Long ago, before the Ice Age, there was no canyon here. Where the gorge now sits, there was only high ground. The Apple River flowed from northwest to southeast; it entered the western end of the park as it does now, but instead of making the sharp turn, it continued down what is now the course of the South Fork, which accounts for the direction of the tributaries and valley widening there. Accordingly, the South Fork now uses the same valley, but it flows in the opposite direction.

What caused the dramatic diversion? The answer lies about 3 miles east. When the Illinoian ice sheet advanced to that position, it blocked the Apple River's flow to the southeast. A lake formed in the dammed river valley and kept backing up to the northwest. At last, the water rose so high that it spilled over the high ground. The resulting rush of water and waterborne glacial rock debris had the effect of a chain saw brought down on soft wood: it rapidly gouged out the canyon. And so the Apple, using this new channel to drain the lake, was diverted southwestward. Later, when the glacier retreated, the river stuck to its new course through the 200-foot-deep gorge, the old valley having been clogged with glacial sediments. This episode is a clear-cut example of how glaciers can profoundly affect their surroundings, even when they do not come into direct contact with them.

Originally, the Apple River flowed from right to left by this cliff of Dunleith formation beds. When dammed by a nearby Illinoian ice sheet, the stream cut a canyon through this rampart. The beginning of the canyon's far wall can be seen in the background.

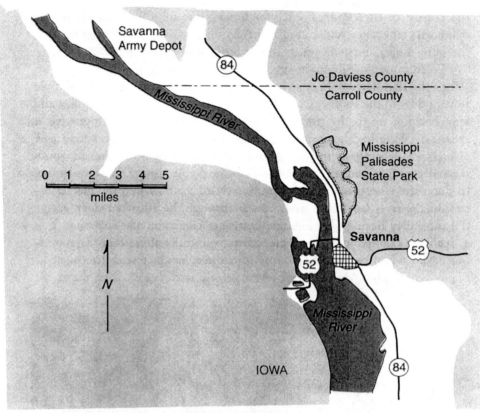

Both entrances to Mississippi Palisades State Park lead to the lofty blufftop, which has several picnic areas and vistas. Proceed to the parking area for Lookout Point, the overlook closest to the southern entrance. The walk to the observation deck is short.

— 3 —

A Scenic Bluff, a Zone of Faults, and the Track of a Mighty River
MISSISSIPPI PALISADES STATE PARK
Carroll County

It has often happened that I have spread acid on the surface of stones which seemed calcareous . . . without producing the effervescence which I awaited.
— Guy de Dolomieu, 1791

For centuries, geologists have used a simple test to determine whether a particular sedimentary rock specimen is limestone. They put a drop or two of dilute hydrochloric acid on the rock's unweathered surface. If the acid fizzes vigorously, it means the rock is calcareous—at least 50 percent calcium carbonate, which makes it limestone indeed. But almost exactly two centuries ago, a French army officer exploring the formations of a range in the Italian Alps began to wonder why the rocks he saw, which otherwise seemed perfect examples of limestone, responded so sluggishly, if at all, to the acid test.

The Frenchman, Dolomieu, was the first scientist to call attention to the new rock type that fizzed so poorly. In the years that followed, other geologists determined that it was primarily composed not of calcium carbonate, but of the closely related magnesium carbonate. To honor the discoverer, they coined the term *dolomite* and affixed it both to the mountain range Dolomieu explored and to the mineral form of magnesium carbonate. The name also came to refer to the rock type, though some geologists now use the term *dolostone*.

One of Illinois's most impressive scenic locales, Mississippi Palisades State Park, is just the place to see a massive outcropping of dolomite formations. The park occupies the eastern bluff of the river north of the pretty waterfront town of Savanna. Approaching the bluff edge on the short walk to the observation deck, take a careful look at the steep-sided, egg-shaped depression in the wooded area west of the parking area. This is a sinkhole, a landscape feature common in limestone and dolomite terranes. Sinkholes form where water gradually dissolves the carbonate rock in a depression or joint exposed at the surface. In time, continued solution and rock collapse cause

Trees growing in the steep-sided bowl of a sinkhole, at Lookout Point. Sinkholes are characteristic of a terrane composed of carbonate rocks.

the opening to become more or less funnel-shaped. The hole, often open at the bottom, may direct rainwater underground, where it promotes the formation of caverns and other solution cavities. Dolomite and limestone do not dissolve in pure water; some carbon dioxide must be present to promote the reaction—and it generally is in our carbon-dioxide-rich world.

When you step out onto the wooden observation deck, you will understand how the sinkhole could have formed. The spectacular cliffs, roughly 250 feet high, are practically all Silurian-period dolomite—with the rather minor exception of the basal 3 feet, which expose a tiny bit of the Upper Ordovician Maquoketa group. To the person who is not an expert in the geology of this area, the task of telling the various dolomite formations apart can be as frustrating as hunting for originality at an Elvis convention. Stratigraphers, however, draw important distinctions based on the fossils present, on the types of weathering, on stratum thickness, and on the texture and color of the rock. These specialists proceed much the same way an art expert does in learning the distinguishing traits of painters from the same era and school.

From this vantage point, the only formations you can see at fairly close range are the uppermost ones. The beds occupying the top 80 feet of the cliff belong to the Middle Silurian Racine and Marcus formations of the

Niagaran series: massive dolomite, which near its base contains a characteristic fossil, the brachiopod *Pentamerus*. Below these lie Lower Silurian, Alexandrian series formations: the Sweeney (45 feet), the Blanding (35 feet), and, just above the base-forming Maquoketa, the Mosalem (10 feet). Niagaran dolomites like the ones at the top of this cliff are extensively quarried in the Chicago area; and it is in the Racine formation in particular that the great Thornton reef (described in essay 15) was largely formed.

What does this preponderance of Silurian dolomite tell us about the Savanna area of some 435 to 425 million years ago? It demonstrates that the region was at the bottom of a saltwater sea, but that is not to say it was a deep ocean basin. Ever since late in Precambrian time, at the very least, Illinois has been a part of the North American continental crust. The long history of the earth records many instances of low-lying continental terrain being submerged by the sea. In modern times, the Grand Banks off Newfoundland are a good example of such submergence. But our continent was in a much different place then than it is now. Geologists who use the plate-tectonics theory to reconstruct the motion of continents through the ages have plotted North America's position in Silurian time. One widely accepted interpretation places it between the equator and 20 degrees south latitude.

The presence of carbonate rocks—most notably, limestone and dolomite—often means that the deposition took place a long distance from the nearest shoreline. (There are notable exceptions, though. Limestone is now forming in shallow water near the Florida Keys, and in other places as well.) In addition, Charles Schuchert, a self-taught paleontologist and stratigrapher and one of America's greatest students of ancient environments, thought

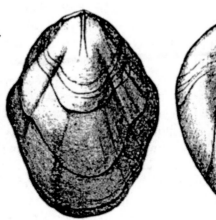

The brachiopod Pentamerus, *Middle Silurian index fossil.*
—Illinois State Geological Survey

that dolomite indicated slightly different conditions than those suggested by limestone. Schuchert (1858–1942) theorized that limestone was the product of wide-open seas in a warm and humid climate, and dolomite was deposited in somewhat shallower, evaporating seas. Other geologists have speculated that dolomite is nothing more than limestone that subsequently was partly altered. Their viewpoint, now accepted by most experts, states that at some point the original limey sediments had their calcium ions replaced, in large or small part, by magnesium ions. Whether this change took place before the sediments turned to rock is a matter of debate.

At this high and breezy overlook, the rock strata in the foreground are not the only thing that attracts the visitor's eye. The river, winding here between narrow bluffs, tells a story as profound as the rocks that define its path. Rivers are immensely important features that have an effect on the earth's surface that would be difficult to overestimate. But how long-lived are they? If you were to guess which last longer—rivers, mountains, or lakes—which would you choose?

Lakes, especially those that are products of the Ice Age, disappear with astonishing speed, in geologic terms. Mountains, often taken as symbols of rugged endurance, are thrown down by the relentless forces of erosion a few million years after their uplift has ceased. As the famous Illinois-born geologist John Wesley Powell noted in his forthright way, "Mountains cannot long remain mountains; they are ephemeral topographic forms." Rivers, on the other hand, can last scores or even hundreds of millions of years. The preeminent geomorphologist William Morris Davis concluded that some of the great rivers on the western flank of the Appalachians have survived more or less intact since the late Paleozoic era. They are about 300 million years old, or half again as ancient as the Atlantic Ocean.

This fact makes us wonder whether the Mississippi—after all, our continent's primary river—is a very old stream, too. In its southern portions, at least, it probably is. Geologists think it likely that a major river followed the Mississippi Embayment—the troughlike structure that reaches north from Louisiana into southernmost Illinois—since the late Paleozoic. The exact course that river took in these northern reaches, in the vast, lost span between the Pennsylvanian period and the last few million years, is difficult to say. Still, it is quite clear that this particular section of channel in front of you may be a mere toddler, which perhaps did not even exist before the Pleistocene epoch. Had you stood here just a little before the onset of the Ice Age, chances are you would have found no Mississippi, and no Mississippi Palisades. You would have been on a low, rolling surface that geologists call the Lancaster peneplain. To the north, in the area of Galena and Dubuque, the tough Silurian rocks on the edge of the Wisconsin arch formed a highland and drainage divide that separated southward-flowing streams from those draining northward.

Starting about 1.6 million years ago, the glaciers came. At an early stage in the Ice Age—before the onset of extensive Illinoian glaciation, some 300,000 years before present—an ice sheet descending from Canada blocked the outlets of the northward-flowing streams, and they backed up into lakes. The lake level eventually rose to the point where the water spilled over the Galena-Dubuque divide and cut a permanent channel through the resistant rock. In the thousands of years of glacial advances and retreats that followed, meltwaters continued to use this new channel to drain to the Gulf of Mexico. But in this initial phase, the Mississippi did not continue south along the valley that now marks the Illinois-Iowa boundary. Instead, it swung south-eastward a few miles south of here, near Fulton, and continued across to almost the center of the state before it returned to its modern valley a little north of St. Louis.

Surveying the scene from this southern rampart of the Wisconsin Driftless Section, note how narrow the valley is here. The river has had to battle its way through these hard, cliff-forming dolomites. A little to the north, in the vicinity of the Savanna Army Depot, and again to the south of downtown Savanna, the valley is up to four times as wide. Can you guess why? In those places the outcropping rock is the Upper Ordovician Maquoketa group,

View of the Mississippi Palisades from Lookout Point. Here the Mississippi River squeezes through a relatively narrow gorge flanked by resistant Silurian dolomites. The valley is much wider just north and south; there the river has cut through the much less resistant Ordovician Maquoketa shale.

predominantly composed of soft shales, which resist sideward erosion about as well as the Chicago Cubs reach the World Series.

Most visitors who leave Mississippi Palisades State Park by heading south on Route 84 note the handsome bridge to Iowa—a favorite roosting place for bald eagles and turkey vultures—but do not realize that just before they reach downtown Savanna, they are passing over a small but significant feature, the Plum River fault zone. Since the mid-nineteenth century, Illinois geologists have recognized that the strata here have a complex geometry. First it was assumed that they formed an anticline, or an upward arch—like a rug that has had its edges pushed together to form a long ridge in the middle: the rug is wrinkled, but the fabric isn't torn. This interpretation of the structure lasted until the painstaking work of State Geological Survey personnel in the 1960s and 1970s revealed that the beds had not just been bent, they had been broken. Those on the southern side of the fault zone had been raised from 100 to 400 feet above the corresponding beds on the northern side. This east-west structure extends about 60 miles from west of Byron, in north-central Illinois, to an area south of Maquoketa, Iowa.

The age of this intriguing if minor rent in the earth's crust can't be precisely determined; we can only say with certitude that it is definitely post-Silurian and pre-Pleistocene—a yawning gap of uncertainty of more than 400 million years. The most reasonable guess dates it from the later part of the Paleozoic era, when the collision of North America and Africa triggered both the Alleghenian mountain-building event to the east, and, apparently, several faults and other stress features that have been mapped in Illinois.

The long legacy of fieldwork and study by Illinois geologists in this locale is a monument to more than patience, careful observation, and sharp insight. It also demonstrates how earth scientists form a theory from an initial set of clues, then refine or replace that theory as more information becomes available. Hence the Savanna anticline becomes the Plum River fault zone. This satisfying process, repeated many times over, often leads specialists and the general public alike to think that science is one steady march toward the ultimate truth. But in each generation, the best practitioners of geology, together with their best colleagues in the other sciences, realize that this is a dangerous assumption. An increase in knowledge is always possible, and always desirable. But we live in a world of such complexity—geological and otherwise—that the exact nature of what was, and what is, will always elude us. Fortunately, most geologists feel right at home here in the Might-Have-Been. Each geologist has his or her private vision of the past that fleshes out the field data.

— 4 —
The Upper Rock River: Part One
LOWDEN STATE PARK
Ogle County

We see on the earth's surface not only the features of the present climate but also those of a past climate. Very extended areas, formerly covered by ice, are now exposed to river action.
—Albrecht Penck, 1905

The locales described in the first three essays apparently were spared from the direct onslaught of the Pleistocene glaciers. The handsome country south and southwest of Rockford can claim no such distinction. It preserves a hilly character that must have made early white settlers, many of them transplanted New Englanders, feel right at home. Even today, the Rock River towns of Byron, Oregon, and Grand Detour have a sense of tidiness and decorum that makes them seem more closely linked to Vermont and Massachusetts than to the plain-faced prairie settlements not far to the south and east.

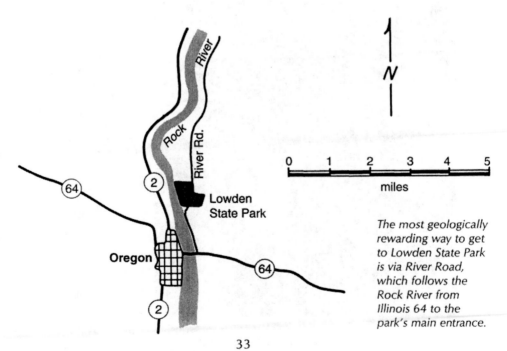

The most geologically rewarding way to get to Lowden State Park is via River Road, which follows the Rock River from Illinois 64 to the park's main entrance.

In Illinois, the physiographic region known as the Rock River Hill Country extends in a southwestward slant from the Wisconsin border east of Rockford to the Mississippi below Savanna. It is a terrain that for the most part was last glaciated in the Pleistocene's Illinoian stage, between 300,000 and 125,000 years ago. In the relatively long span that has elapsed since the last retreat of the Illinoian ice, the drainage system has removed some of the ice-borne deposits and has reaccentuated the rolling surface that had developed in preglacial times. If you drive through the farmland south of Mount Carroll, for instance, you will find a landscape reminiscent of the Wisconsin Driftless Section.

This essay and the next are devoted to the single most beautiful slice of the region: the geologically complex upper Rock River valley. The valley contains a string of geologic wonders, all to be found in the space of one short, scenic drive. A good starting point for your exploration is Lowden State Park, a favorite site for campers and picnickers, located on the river's eastern bluff, north of Oregon's town center.

Just before you turn left into the park when approaching from the south, the road climbs up the bluff through a section of roadcuts. These expose narrow-bedded dolomite layers belonging to the Ordovician Platteville group, which in the Galena area has proven to be the primary zinc-ore zone.

Thin-bedded layers of Ordovician Platteville group dolomite stand guard along the road on the approach to Lowden State Park. Lens cap for scale.

34

This glacial erratic, a highly decorated visitor from the Canadian Shield, was transported to northern Illinois from the region of Precambrian outcrops. Note the horizontal quartz vein in the dark, finely crystalline xenolith to the right of the ornamental plaque.

Not far from this outcrop is a small quarry face where the oldest surface rocks in Illinois are exposed. They belong to the Cambrian Potosi and Franconia formations. Unfortunately, the quarry is on private land. Your best bet for seeing it is to take the next Illinois State Geological Survey field trip held in the Oregon area. In the meantime, you can take heart in the knowledge that there is an even more ancient sight to behold in Lowden State Park. Not far inside the entrance, on the northern side of the roadway, stands a glacial erratic that was carried from its home in the Canadian Shield by one of the Pleistocene ice sheets. The rock bears a bronze plaque honoring former governor Frank Lowden. This boulder is a particularly striking igneous specimen. Examine the fine assemblage of crystals in the handsome white granite and note how the lighter quartz and feldspar minerals are offset by flecks of the darker hornblende and biotite. Within the granite, you will also see another igneous rock type: one that resembles, at first glance, the dark-toned diabase. It is a xenolith, a foreign rock body, that became embedded in the granite while the latter was still molten. This erratic is Precambrian in age; like the many other crystalline rocks that hopped a ride with the glaciers, it was originally part of formations that were created hundreds of millions of years before the Cambrian and Ordovician sedimentary strata of this locale were deposited.

The view west across the Rock River, from the American Indian statue site. Route 2, the main roadway on the far side, runs up a terrace of glacial outwash.

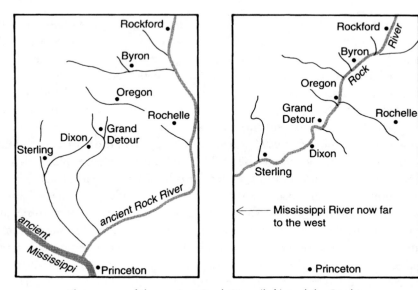

The course of the ancient Rock River (left) and the Rock's modern course (right). —Illinois State Geological Survey

In Canada and northern Wisconsin, Precambrian outcrops are a common sight. But what many people do not realize is that this ancient crystalline bedrock extends hundreds of miles to the south, though buried far beneath the surface. To see it in Illinois, you'd have to do a great deal of digging. In this area, for instance, it is covered by about half a mile of Paleozoic sedimentary strata; if you try farther to the southeast, in the heart of the Illinois Basin, you'd have to go straight down seven times as far. So to explore the Precambrian in the Land of Lincoln, you might as well stick to the wide variety of igneous and metamorphic erratics that grace parking lots and driveways from one end of the state to the other.

The main venue at Lowden State Park is the river vista that stretches in front of the massive statue honoring the American Indian. After you have examined the statue's weathered concrete exterior, glance up and down the stream course. This lovely setting through which the Rock now rolls is, geologically speaking, one of the most interesting segments of the valley. Up to the late Pleistocene, the river flowed southward from the Rockford area roughly 10 miles east of here, and about 25 miles east of modern-day Dixon. Finally, it joined the easternmost segment of the ancient Mississippi near the Bureau County town of Princeton. When runoff from early Wisconsinan glaciers clogged its valley with drift, the upper Rock backed up north of here. The ponded water finally spilled over a divide between Byron and Oregon and cut this narrow defile.

Now examine the far side. A string of automobiles zipping north and south should make it easy for you to spot Route 2, which runs across a high terrace, about 45 feet above river level. This terrace is formed from valley train deposits—river-borne glacial sediments that choked this waterway in late Wisconsinan time. You should be able to discern the semblance of a lower terrace, too, about halfway down to the river. This level marks an even later spate of valley train deposition, which took place after the Rock had resumed its downcutting.

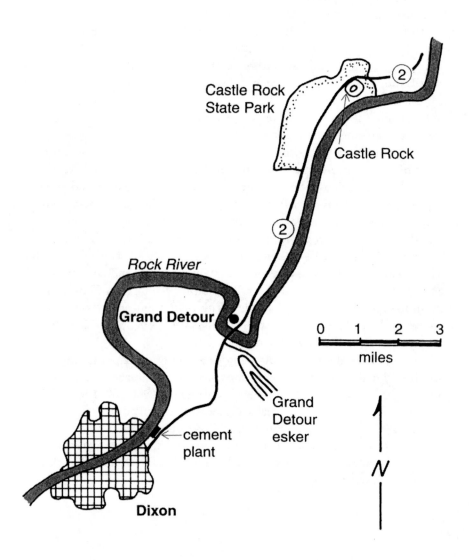

Castle Rock State Park and environs.

— 5 —
The Upper Rock River: Part Two
CASTLE ROCK STATE PARK TO DIXON
Ogle and Lee Counties

I perceive that the surface of the earth was from of old entirely filled and covered over in its level plains by the salt waters.
—Leonardo da Vinci (1452–1519)

No other vantage point on the Rock River matches the observation deck at Castle Rock State Park. There, in addition to the sweeping treetop panorama, the visitor gets a close look at one of Illinois's most beautiful and most economically exploited rock formations. This exposed bedrock is a testament to Leonardo's observation that what is now dry land was once a completely different, marine environment.

When you reach the park's wooded confines, turn into the gravel lot on the east side of Route 2, just south of the prominence that gives the park its name. There are two overlooks, and the best one is up the long set of wooden steps to the north. It takes you to the summit of Castle Rock, where the prospect is breathtaking. It is a good reeducation zone for any out-of-stater who believes Illinois is scenically depauperate.

The cliffs you view from the observation deck are held up by one of the Midwest's most instantly identifiable rocks: the crumbly and crossbedded St. Peter sandstone, of Middle Ordovician age. It often bears brown or rusty tints from the weathering of one of its associated iron minerals, limonite. But when it shows its pure white face along a newly exposed surface, it can almost be mistaken for a white-sugar confection. If you scrutinize it with a hand lens or a magnifying glass, it reveals a different kind of beauty on a much smaller scale. It's composed of frosted, rounded, and evenly sized grains of light-toned quartz. This kind of quartz is highly prized as the source of exceptionally pure silica used in glassmaking and a host of other industrial applications.

You should definitely examine the St. Peter, but do not collect specimens here. Also, heed the signs asking you not to damage the fragile exposures by wandering off the pathway. In flat places, you'll see that some of the sandstone has been stripped of its weak cementing matrix and has reverted to unconsolidated sand; after serving as rock for almost half a

39

The final approach to the Castle Rock summit. Weathering and erosion are slowly causing the St. Peter sandstone to revert to its ancient form—loose, uncemented sand.

billion years, it has returned to its original form. It offers microhabitats for plants more commonly encountered on the windy dunes of Lake Michigan. The last time these opaque grains of silica were loose, there was little or no life on the harsh, stony, ultraviolet-drenched surface of the earth. Illinois lay a little south of the equator, in a framework of continents no one today would recognize.

Try to picture the environment in which the St. Peter sandstone was laid down. Was it sea or land? No one disputes that the sand originally came from the disintegration of still older rocks of the Canadian Shield. But into what kind of conditions was the sand deposited? For a long while, geologists speculated that the St. Peter was the product of a desert environment. After all, it exhibits crossbedding patterns, a characteristic of dunes. Close examination of the crossbedding suggests that the sand was deposited in the sea instead—though close to shore, where energetic surf or tidal action could produce the grains' typically well-rounded shape. A huge amount of sand was laid down in the making of this formation. The St. Peter extends like a thick, gigantic skirt flung over the long edge

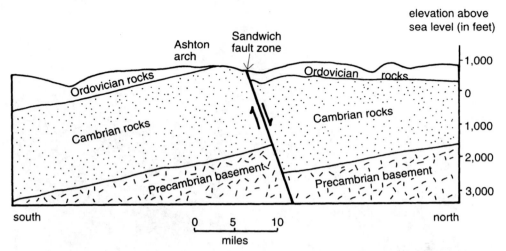

Simplified cross section of crustal displacement along the Sandwich fault zone. —Illinois State Geological Survey

of the continent's core. It covers thousands of square miles, at the surface or below, in America's midsection.

Less obvious from this observation deck, but no less geologically significant, is the fact that you are standing in the midst of one of northern Illinois's most structurally complicated areas. Running just about under your feet, northwest-to-southeast, is the axis of the Sandwich fault zone. Actually a group of faults, it can be visualized, not all that inaccurately, as one overall normal fault—where one side, the so-called hanging wall, has slipped down at an angle between 45 and 90 degrees to the horizontal. Were you able to cut open the earth here with a giant knife stroke and see the fault in cross section, you'd find that the two sides have moved the way gravity would seem to dictate. For that reason, normal faults are sometimes called gravity faults.

The north side of this fault is the one that has moved downward; the maximum vertical offset is reportedly about 900 feet. Paralleling the Sandwich fault zone, on its southern flank, is the Ashton arch, a major anticline that exposes some of the state's most ancient rocks, from the first period of the Paleozoic era, the Cambrian. And a little north of the Sandwich fault zone lies the Oregon anticline, another elongated arch structure, which brought the Cambrian rocks described in the previous essay to the surface near Lowden State Park. The active phases of the Sandwich fault system and its associated anticlines seem to have been concentrated in two pulses, at the end of the Mississippian period (about 320 million years ago) and at the end of the Pennsylvanian period (about 286 million years ago).

41

A small borrow pit excavated into the Grand Detour esker. Eskers are a premium source of clean, well-sorted sands and gravels. The sediments exposed in this cut were originally deposited in a tunnel or long groove in a stagnant Pleistocene glacier.

On leaving Castle Rock, head south on Route 2. You'll soon reach the handsome little settlement of Grand Detour, the home of Vermont-born John Deere and his famous stainless-steel plow. The settlement sits chastely on the northern side of a dramatic bend in the Rock River. The stream's contorted course first heads southwest, then northwest, then northeast, then west, then south, then southeast, then southwest again. One of the main reasons the river has executed this loop can be found on the southern side of the Route 2 bridge. If at the first side road (Detour Road) you turn either right or left, you will be paralleling what has been described as the Grand Detour esker, a ridge of Pleistocene sand and gravel that apparently presented enough of a barrier to make the Rock swing around its far end. This esker, which in recent times has borne the brunt of quarrying, road building, and residential construction, can be traced southeastward some distance, but it can be difficult to distinguish from till-covered bluffs and deposits of wind-blown sand. The Grand Detour esker is an enigmatic feature. Originally geologists thought that the Wisconsinan ice sheet extended westward from the Lake Michigan basin into this area. Recent interpretations, however, state that the glacier stopped short of this locale.

According to the prevailing theory, eskers form when glacial meltwater runs through tunnels or grooves in the ice, near the margin of a stagnant glacier. The tunnels eventually become filled with coarse and well-washed sediments. When the ice melts away, a positive landform emerges.

The Dixon-Marquette cement plant, on the north side of Dixon, depends on a reliable source of lime derived from carbonate rocks.

The Rock's detour may have been partly the result of the surface expression of another of the area's zones of deformation. This feature, the La Salle anticlinorium, is the longest structural feature of its kind in the state. It runs from Ogle County all the way down to southwestern Indiana. The anticlinorium is a complex of strata arched upward into anticlines and, in the intervening troughs, downward into synclines. Unless you are a geologist with the time and skill to decipher the area's well-core records, the chances are that you won't notice this set of ripples in the crust. While it has had a defining effect on which rock types outcrop where—in an anticlinal structure, older beds are on the arch's crest and younger rocks are on its flanks—it isn't something most people can recognize easily. The La Salle anticlinorium apparently has a history of deformation similar to that of the Sandwich fault zone discussed above.

A minute or so after you've passed over the anticlinorium's invisible crest, you are in the city of Dixon. This big, industrialized river town defines itself as the state's Petunia Capital. But from the perspective of the geologist, the architect, or the civil engineer, it might as well be rechristened Cement City. It is one of Illinois's big production centers for the construction material that literally holds the skin of modern civilization together. The most prominent landmark along Route 2 north of downtown is the Dixon-Marquette cement plant and its associated quarry pit.

Cement is the binder used in concrete and other construction applications. Nowadays, the most commonly used variety is Portland cement, invented in the early 1800s by the Englishman Joseph Aspdin. Its name is taken from England's Portland architectural stone, which, it is said, it resembles.

Carbonate rock (limestone or dolomite) is a crucial component in the making of cement. Production begins when lime is extracted from the stone and mixed with clay, shale, or foundry slag. The mixture is burned in kilns, and the resulting ash, known as clinker, is ground to the extremely fine consistency needed for its various applications. The effectiveness of cement is due to the unusual nature of the calcium silicates it contains. When they are added to water, they radically change their nature and develop the interlocking crystals that give cement its binding power.

— 6 —
A World of Wind and Melting Glaciers
THE PRINCETON AREA AND THE GREEN RIVER LOWLAND
Bureau, Henry, and Lee Counties

Ye living ones, ye are fools indeed
Who do not know the ways of the wind.
—Edgar Lee Masters,
Spoon River Anthology

Here in the heartland of North America, the ways of the wind are many. The ever-moving cushion of air brings a ceaseless procession of tornados and ice storms, brisk autumn breezes and soaking spring rains. Even so, few Illinois residents fully realize how much the force of the wind has shaped and reshaped their state. Sand dunes, those quintessential aeolian features, may seem exotic productions of dry desert regions; but several sections of Illinois, including one part of its northwestern quarter, are dotted with them.

However effectively the wind shapes dunes, it cannot create their raw material. Sand, the sediment required, is usually carried and deposited by river systems. In the case of the region known as the Green River Lowland—most of which lies in Lee, Bureau, Whiteside, and Henry Counties—the sand was delivered by a much more impressive transportation system, the ice sheet of the Wisconsinan glaciation. It is this interaction between sand and glacial ice that we'll investigate in this relatively unknown but fascinating part of the Prairie State.

Our first local Ice Age benchmark is in the Princeton area of Bureau County. It's virtually impossible to recognize the fact at first glance, but the town of Princeton, located in the physiographic section known as the Bloomington Ridged Plain, stands atop a much older landform, the buried junction of the bedrock valleys of the ancient Rock and ancient Mississippi Rivers. The old Rock stream course, dubbed the Paw Paw Valley by geologists, extends northeastward; that of the old Mississippi, appropriately called the Princeton Valley, reaches northwestward. During the Pleistocene epoch's concluding Wisconsinan stage, the far-reaching Shelbyville glacier pushed both the Mississippi and Rock far westward, to their present locations. Other, later glacial advances, including the Bloomington, entered the area and finished filling in the abandoned river valleys.

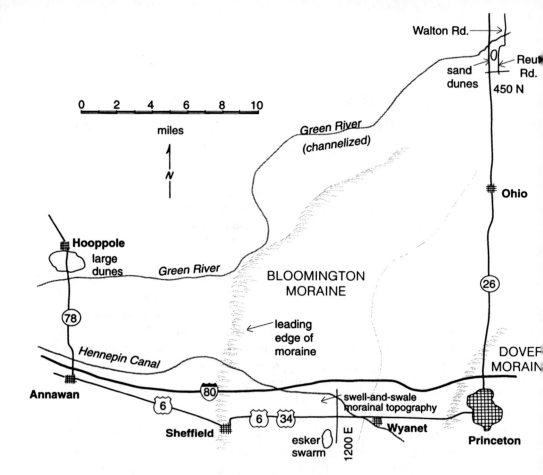

The geologic tour of the Bloomington moraine and Green River Lowland begins in Princeton, south of Interstate 80.

You can see superb examples of Ice Age landforms just west of the small town of Wyanet, which is approximately 5 miles west of Princeton on Route 6/34. This route will take you up the low, rolling backside of the Bloomington end moraine (this segment of the feature is also known more specifically as the Providence moraine). You will cross over the historic Hennepin Canal, which once linked the Illinois River with the upper Mississippi via the Rock River. Two miles west of Wyanet, turn south onto the gravel road 1200 East and continue for another 1.5 to 2 miles. Keep a sharp lookout for low, sinuous ridges running across the surrounding terrain. You are in a lovely swarm of eskers. These serpentine landforms were created near the edge of the Bloomington glacier when it became stagnant; had it still been moving forward, it would have oblit-

The low ridge in the foreground (dotted with scattered trees and shrubs) is a superb example of an esker. You can visualize the glacial tunnel in which the sediments were deposited. Located in an esker field to the west of Wyanet.

erated them. Meltwater coursing down tunnels in the glacier—or perhaps in channels on top of it—filled with layers of sand and gravel. When the ice melted away, the sediments remained, as a sort of mirror image of the original stream channels.

If you take 1200 East back up to the Route 6/34 junction and continue straight north on it, you will have a good opportunity to examine the characteristic look of an end moraine's upper surface. You'll quickly sense that it is anything but flat; there are hummocks and hollows. Geologists call this swell-and-swale, or knob-and-kettle, terrain. Wherever you go in glaciated Illinois, end moraines have this undulous aspect. Can you guess what lies directly underneath the uneven surface? Is the moraine composed of hard bedrock, scoured and sculpted by the relentless ice; or is there softer material—loose sediments that had been carried by the glacier and deposited at its edge?

In fact, end moraines are composed of a mishmash of unconsolidated sediments collectively termed till. Till is poorly sorted—it contains particles of widely varying shapes and sizes, from clay and silt to cobbles and even boulders. In contrast, the sand of dunes and the silty loess found on many river bluffs are well sorted by the wind, so that only one particle size predominates. The till of the Bloomington moraine is well known for its telltale coloration. As you explore the Wyanet area, you may come across a fresh

Black oaks on the flank of a sand dune, just south of the Green River.

roadcut. If you do, see if you can detect the characteristic pink tint of the Bloomington drift. (For more on how end moraines form, see essay 25).

The glaciers of the Princeton and Wyanet area did more than build end moraines and eskers. In filling up the bedrock valley of the ancient Mississippi, they created a lowland that eventually received a huge amount of meltwater and outwash. But what exactly is outwash? Essentially, it is material found in one of three circumstances: it was deposited in meltwater channels leading away from the glacier (in this case, it is called valley train); it takes the form of layered sediments in ice-contact features such as eskers and kames; or it is an outwash plain, a broad, gently sloping sheet of sand and gravel extending away from an end moraine—as is the case here. To see this large outwash plain, the Green River Lowland, either travel westward on Route 6 toward Sheffield and Annawan or backtrack to Princeton and head north on Route 26 toward Ohio (the town, that is) and Dixon. The first alternative leads you to the dune-studded lowland at its maximum extent. If you take this leg, I suggest you turn north on Route 78 when you reach Annawan. As you cross the Green River and approach the Hooppole area, you will see some hefty dunes—most of them under cultivation—in the surrounding farm fields. You will also get a good feel for the enormous volume of outwash that created this broad expanse.

The second option is my preferred route, because it takes you over the overlapping Providence, Buda, and Sheffield sections of the Bloomington end moraine on the way to the narrower, eastward extension of the Green River Lowland. To me, a native of moraine-studded Lake County, it's almost as good as being home again. As you descend the front face of the Bloomington rise north of Ohio, turn east on 450 North, then turn left on Reuter Road, which leads you to the Green River floodplain. Before you swing east and north onto the Walton Road/1000 East bridge over the Green, however, you enter a zone of humpy, lumpy land covered with sand-loving black oaks. This is another of the lowland's many clusters of sand dunes. (Much of the land here is privately owned; take pains to avoid trespassing.) Farther north, as you near the Green River, you can pause for a moment by the roadside and ponder the forces that brought these dunes into being.

Picture yourself standing here a little more than fifteen thousand years ago. The once-mighty Bloomington glacier, which originally had marched down the Lake Michigan trough and spread like an immense amoeba over northern Illinois, is now in full retreat a few miles to the south. Its meltwaters rush over the plain, depositing layer after layer of outwash. In the years that follow, the plain is one vast surface of blowing and drifting sediment—a huge beach without a shore. Vegetation has not yet arrived in sufficient quantity to anchor the loose material. The sand, just light enough to be hefted short distances by the winds, accumulates in pockets here and there. These accumulations form obstructions that momentarily reduce the wind's speed; so more sand falls on them. Thus are dunes born. At first they are anything but static features. They actively migrate as the prevailing winds direct, until plants arrive and hold them fast in one location.

Scientists who study the complex subject of sedimentation classify dunes by shape. One highly respected text on the subject lists no fewer than seven different dune types. A dune field may have one prevailing type—for instance, the great crescent-shaped barchan form of Colorado's Great Sand Dunes National Monument—or it may contain a chaotic jumble. In the Green River Lowland, the majority of dunes are the parabolic (or U-shaped) type. Like barchans, they have two projecting points; but the points face upwind. You may also come across longitudinal dunes. They resemble long ridges that lie parallel to the prevailing wind direction.

If you cross the Green on Walton Road/1000 East, you'll see another geological phenomenon: a much more recent one, directly caused by *Homo sapiens*. The river has been straightened, or channelized, for much of its length. Whenever the distance between a stream's head and mouth is shortened, certain changes take place: water velocity increases, and downcutting becomes more pronounced. Accordingly, the river entrenches itself and carves steep banks. You'll see evidence of that if you stop at the bridge and take a look upstream and down.

North of the river, you can find more clues indicating the presence of glacial outwash. Huge circle-pivot irrigation rigs, standing like giant lawn sprinklers escaped from a sci-fi nightmare, loom above the farm fields. These are typically used to water crops growing in well-drained, sandy soil. You will also catch sight of tidy-looking patches of pine trees on carefully tended plantations. The southernmost relict population of wild white pines, in contrast, is located miles to the north, in a state park north of Dixon. These trees were planted as a paying proposition. In the eye of the passing motorist, they are one of the most pleasing agricultural uses of sandy soil.

Huge circle-pivot irrigation systems often indicate farmland in areas of glacial outwash. Crops growing in this kind of sandy, well-drained soil need extra water. A heavy rainfall created the temporary pond in the foreground a few hours before the photo was taken.

— 7 —

Cliffs, Canyons, and Catastrophes: The Upper Illinois River

BUFFALO ROCK AND STARVED ROCK STATE PARKS

La Salle County

It is only the philosopher, who has deeply meditated on the effects of which action long continued is able to produce . . . who sees in this nothing but the gradual working of a stream, that once flowed as high as the top of the ridge which it now so deeply intersects.
—John Playfair, *Illustrations of the Huttonian Theory of the Earth*, 1802

At the end of the eighteenth century and throughout much of the nineteenth, a succession of three Scotsmen played a crucial role in the development of the earth sciences. This geological trinity—James Hutton the creative theorist, and John Playfair and Charles Lyell his talented proponents—persuasively argued for a vision of the earth now known by the cumbersome name of uniformitarianism. One of the many potent ideas contained in this theory is that small effects sufficiently prolonged can produce immense results. In fact, the Scottish trio sought to prove that the majority of geological processes fall into this category: the earth is largely a perpetual-motion machine of gradual change endlessly compounded. In this view, catastrophes can happen—but only if they are neither more widespread nor more powerful than processes we can observe today.

This concept led to one of the greatest realizations in the history of human thought: given the number of changes unequivocally recorded in the rocks, the earth must be very, very old indeed. Not only was this proposition unsettling; it was often accurate. Had it not been for Hutton and his able propagandists, we might still be trying to shoehorn all the earth's eras into the span we now reserve for a single epoch.

Uniformitarianism held the field practically unchallenged for generations—even into the early 1970s, when the plate-tectonics revolution was well under way. Today, however, after two hundred years of fruitful exploration and research, we have incontrovertible proof that the world is a complex interaction between the slow working of subtle changes and the massive, even worldwide changes wrought by short-term, catastrophic events.

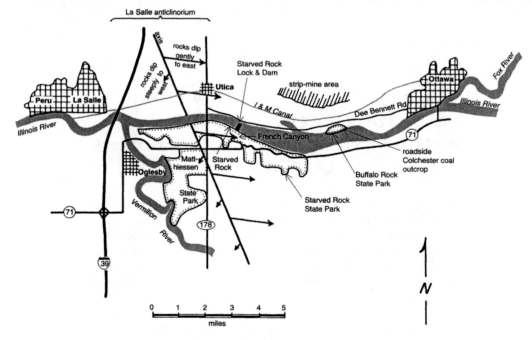

The upper Illinois River valley in La Salle County.
Note the location of the La Salle anticlinorium.

In the locale this essay covers, you'll come across the results of both processes: the patient, centuries–long work of stream action, and the swift but relatively short onslaught of raging floods. Several state parks and natural areas located amid the intense industrial hubbub of the upper Illinois River valley illustrate these themes. The two state parks described here, Buffalo Rock State Park and Starved Rock State Park, are among the best of the lot, though Matthiessen State Park has both exceptional canyons cut into the St. Peter sandstone and massive bluff outcrops of the Pennsylvanian strata that overlie Ordovician formations.

The first main attraction, Buffalo Rock State Park, is worth an unhurried visit, because it has accessible outcrops that contain rocks from the two Paleozoic periods represented in this locale: the Ordovician and the Pennsylvanian. But before you swing off Dee Bennett Road west of Ottawa to enter the park, take a good look at two man-made features that do much to define the landscape of the upper Illinois River area. The first and much more obvious is the line of strip-mine spoil piles that march along the river's northern bluff. The seam of Colchester coal sits close to the surface here, which encouraged mining companies of the twentieth century to strip off the thin blanket of overburden rather than use the standard nineteenth-century method of mine shafts and underground extraction. The effect on

52

the landscape has been great and often gruesome. Laws enacted in recent decades require that mining concerns reclaim areas they disturb; but the large tracts strip-mined by earlier generations remain much the way they were after the companies moved on to greener pastures—or rather, to remaining coal deposits elsewhere. The other feature that may whet your curiosity runs along Dee Bennett Road some distance from the base of the despoiled bluff. The old Illinois and Michigan Canal is a narrow and now bargeless waterway recently designated a National Heritage Corridor. It had a cultural and commercial impact in the nineteenth century that far outstrips its humble aspect. (For more on the creation and significance of the I&M Canal, see essay 13.)

Entering Buffalo Rock State Park, follow the access road as it swings around the high mass of layered sedimentary rock. Take note of the highest section, located along the road at the entrance to the parking area. Before you take in the river view, return to the high part of the outcrop and scrutinize it carefully. This is the top-to-bottom sequence: about 20 feet of gray shale; then about 18 inches of the much-sought-after Colchester coal (sometimes still referred to as the No. 2 coal); and then a narrow, brown strip of underclay, the material usually found directly under coal seams. So far, all this has been Middle Pennsylvanian in age. The rest of the outcrop consists of the famous Middle Ordovician St. Peter sandstone, which with its frosted white quartz grains is instantly recognizable to anyone who has paid a visit to Ogle County's Castle Rock (see essay 5).

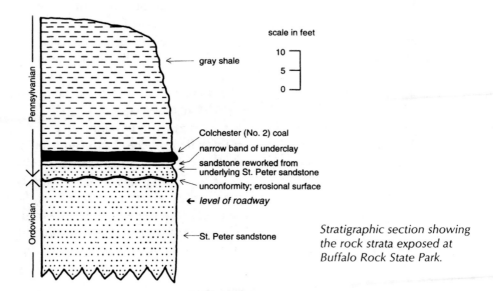

Stratigraphic section showing the rock strata exposed at Buffalo Rock State Park.

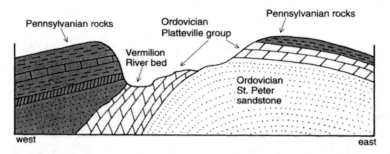

Simplified cross section showing the La Salle anticlinorium in the vicinity of Matthiessen State Park. —Illinois State Geological Survey

When the underclay formed as the soil layer for the luxuriant Pennsylvanian vegetation, the St. Peter had already been rock for some 170 million years. In fact, the sandstone was the bedrock surface on which this coal-swamp forest established itself. You might wonder what became of all the Silurian, Devonian, and Mississippian strata belonging to the time between the Ordovician and the Pennsylvanian. There are two possible answers: either no intervening sections formed at all—because this area was above sea level at the time—or they were deposited, but subsequently removed by erosion. The evidence suggests that the second explanation is correct. After all, relatively undisturbed Silurian and Devonian strata are still present at the surface or underground in large areas both to the east and to the west. So they may have existed here, too, only to be eroded by local uplift triggered, at least in part, by the creation of the La Salle anticlinorium.

Walk from the main parking lot along the bluffside trail until you reach the observation deck, which offers an excellent view of the Illinois River. Do you recognize the rock you're standing on? It's the good old St. Peter again. This formation supplies the region's important silica-based industry with an extremely pure quartz that is the envy of the glassmaking world. The city of Ottawa, just upstream, is the industry's de facto capital; it might as well be renamed Silicopolis.

The sand quarried from the St. Peter has other important applications: as steel-foundry molding sand; as an ingredient in the fracture treatment that increases oil-well productivity; and as a component in abrasives, enamels and other paints, porcelain, and tiles. Because the sandstone is often poorly cemented together, quarry workers can usually extract the sand with nothing more than a stream of water from a high-pressure hose. The jackhammer- and dynamite-toting workers of Vermont's granite quarries would be greenly envious of how easily this Prairie State bedrock gives up its treasure to those who seek it.

54

Looking eastward up the Illinois River, at Buffalo Rock State Park. The cliff in the foreground is Middle Ordovician St. Peter sandstone. Its upper surface weathers unevenly to produce bumps and hollows.

At river's edge, the sandstone surface assumes peculiar shapes and patterns that reflect its varying amounts of iron-oxide cement. Where the cementation is stronger, the rock has resisted erosion more. There are also places where the rock displays fluting structures, showing where waters much higher and much faster than those of the present Illinois have scoured it. This evidence of powerful erosional forces leads us to consider the turbulent conditions that created the river in the first place.

As noted in essay 3, rivers can be very old landforms; but this upper stretch of the Illinois is a mere pup. It is even younger than our own fledgling species—a true sign of geologic youthfulness. The first chapter in the evolution of the upper Illinois began just yesterday, geologically speaking, in the Pleistocene epoch's final, Wisconsinan glaciation. Some 20,000 years ago, the Shelbyville glacial advance pushed the ancient Mississippi River, which originally had flowed by the sites of Princeton and Beardstown, to the more western, state-boundary valley we know today. Approximately 4,500 years later, an episode of catastrophic flooding called the Kankakee Torrent was triggered when meltwater lakes impounded by moraines north and east of here burst through their barriers and coursed down this channel. All hell broke loose. The amount of water escaping the upper Midwest through this one area would have been a staggering sight to anyone accustomed to today's peaceful setting. (The mighty St. Lawrence River, which in modern times is the Great Lakes' main outlet, was completely blocked by ice still present farther north and east.) In at least one location, northeastern Illinois received meltwaters not only from the Lake Michigan ice lobe but also from the more eastern Saginaw lobe. According to recent research by geologist Edwin Hajic, there is evidence of this flood's great scouring power in the surrounding uplands, up to 640 feet above sea level—about 180 feet above the current level of the Illinois River.

According to Hajic's research, the relatively narrow river valley before you was the Kankakee Torrent's inner channel. Buffalo Rock, and Starved Rock across the river, were not examples of a terrace level cut by a later, lower flood, as was previously postulated. In fact, later floods, caused by the periodic draining of Lake Chicago and Lake Nipissing (ancient forerunners of Lake Michigan), did unquestionably occur. But in Hajic's model, Buffalo and Starved Rocks were formed very early on, as "erosional residuals"—isolated masses of bedrock that, at least in part, survived the almost unimaginable erosive surge of the Kankakee Torrent. Later, from about 14,000 to a mere 4,000 years ago, the periodic Lake Chicago–Nipissing floods arrived, and they deepened the river's bed even more. During these spates of rapid downcutting, the tributaries of the Illinois were left hanging at their former level. This resulted in the eventual formation of many canyons and waterfalls, features you'll see in abundance at the next stop.

To anyone born and raised in Illinois, a grade-school field trip to Starved Rock State Park is almost unavoidable. Few are the modern residents of Illinois who have not heard the grisly tale of the eighteenth-century Illiniwek warriors who, besieged at Starved Rock by their Potawotami enemies, chose starvation over surrender. What is not so well known is that American Indians inhabited this area at least 10,000 years before present, in the waning years of the Pleistocene.

Most visitors begin their exploration of the park by making the easy climb to the top of Starved Rock. The cliffs of St. Peter sandstone are dappled by liverworts, thought by many biologists to be some of the most primitive members of the plant kingdom. Indirect evidence, in the form of fossil spores found elsewhere in the world, hints that the ancestors of these flat-lying, stemless organisms may have been the first plants to colonize the land. If this theory is true, it means the greening of the earth's surface began in the Ordovician—the same period that saw the creation of this sandstone—

rather than in the following Silurian, which is still the official genesis of land plants cited by most paleobotanists.

And so on to the canyons. The most accessible from the Starved Rock footpath is French Canyon. One of the loveliest times to see it is in winter, when the falling streams have frozen into sparkling columns of ice. To me, these narrow rockbound defiles, with their horizontal ribbing of Paleozoic strata, define the Illinois landscape every bit as much as the prairies. What gives the canyons of this park their special secluded dignity is the fact that the St. Peter formation does not erode uniformly: when the tributary meets a more solidly cemented rock zone, it perches there and spills down, all at once, to its much lower bed. The waterfall is an expression of the stream's attempt to readjust itself to the modern base level. Slowly but surely, despite the tougher rock, it excavates its way upstream in classic uniformitarian fashion. In the meantime, it has a curiously soothing effect on the soul. Illinois is not just a flat, cloud-dappled expanse; from one tip to the other it is a place of unexpected twilit recesses, secret windings through layered rock, and the plashing sounds of dripping water.

Frozen cataracts at Starved Rock State Park in winter. This small subsidiary waterfall, with its many larger companions, is slowly but inexorably cutting back the walls of French Canyon.

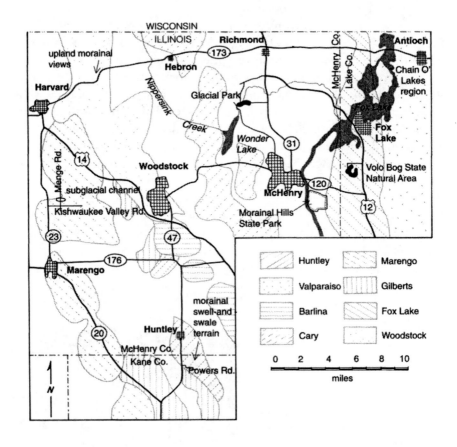

Site map and moraines of McHenry County and environs.

8

A Land Like the Rolling Sea
GLACIAL PARK AND ENVIRONS
McHenry County

The author believes that . . . ice once reached clear to the southernmost edge of the district which is now covered by those rock remnants; that this, in the course of thousands of years, gradually melted back to its present extent; that, therefore, those northern deposits . . . are nothing other than the moraines which that enormous ice sea left behind on its gradual withdrawal.
— Reinhard Bernhardi, 1832

On a lovely summer's day in northeastern Illinois, when hawks circle and the lake breeze ruffles the rounded masses of the farmyard oaks, it is difficult to imagine that the glaciers once came here. The advent of a vast continental ice sheet, in these parts perhaps 2,000 feet thick, simply does not seem possible. Yet this region is one of the best places in the world to see the profoundest effects of the Pleistocene. Elsewhere in the state, end moraines and other glacial landforms are subtle enough to be overlooked by most people; but in McHenry County the landscape rises and rolls and falls away with the grandness of an ocean worked with deep swells. If there is one best place to encounter Ice Age Illinois, this must be it.

Were you to ask one hundred geology undergrads who it was who first formulated the concept of extensive continental glaciation, at least ninety-nine would answer that it was the eminent Swiss-born scientist Louis Agassiz. In 1840, Agassiz published the profoundly influential *Études sur les glaciers,* suggesting that at some point in prehistory the earth's ice sheets had not been confined only to polar and mountainous regions. Agassiz—a man who spent the bulk of his long career not pondering ice ages but poring over fossil fish—was not the only person to propose an earlier time of intense cold. The lesser-known Bernhardi, quoted above, reached the same conclusion in print eight years before; and earlier still, Goethe, Germany's great man of letters (whose achievements as a scientist are now too often overlooked), speculated about an earlier age of widespread ice. It was a concept screaming to be born.

The main focus of this essay is Glacial Park, one of the state's many superb county-run preserves that serve a wide variety of constituencies:

Map of
Glacial Park.

bird–watchers, hikers, snowmobilers, and horseback riders. Earth-science enthusiasts may never constitute a majority in any community; but here they should have an intense proprietary interest. This site is one of the state's most important geologic treasures. It is also a place of uncommon beauty, even by this beautiful region's standards. Visitors sense this at once: the place is at peace. It seems exempt from the chaos and clutter that too often infest our lives.

However you approach Glacial Park, you'll get a good sampling of McHenry County's glorious morainal terrain. The park is situated between the high, scenic ground of the Cary and Fox Lake end moraines. These north-and-south-trending ridges are but two of several such features that locally define the zones where the shrinking Wisconsinan glacier staged minor readvances, then paused long enough to dump the sand, gravel, and till it carried in long, usually continuous bands. The Cary moraine forms the impressive wooded heights west of Wonder Lake; the Fox Lake moraine, a subdued, hummocky area that is actually lower than the outwash plain beyond it, runs along the western side of a body of water that shares its name. In between, in the outwash plain, winds the eastward-flowing Nippersink Creek. (For more on how end moraines form, see essay 25.)

I recommend you start your exploration of Glacial Park from the pulloff on Keystone Road, just east of the intersection with Barnard Mill Road. Pay your respects to the sparkling little Nippersink before you come upon the main attractions. As you walk along this small stream, consider its position

60

in the landscape. Does it occupy a narrow valley, as befits its modest size, or does it flow across a large open area?

It doesn't take a careful study of topographic maps or a hawk's-eye view to see that it does the latter. The phenomenon of wide valleys occupied by tiny streams—geologists call them underfit streams—is a fairly common one in the Prairie State. The modest creek you see here today is a mere trickle in comparison with the waterway that once helped drain the surging meltwaters of the Wisconsinan ice sheet. And the Nippersink is also flowing in the wrong direction, at least in terms of what took place here in the late Pleistocene. At that point, the lowlands to the east were still blocked with ice; consequently, the drainage was oriented westward. The local tributaries did not flow the way they do now until the region's master river, the Fox, settled on its more eastern valley.

Not long after you cross the footbridge over the Nippersink, you'll get an unobstructed view of the park's chief point of interest. It is a sight liable to make a geomorphologist of advanced years do handsprings. Rising up in front of you, like the huge humped back of a Jules Verne sea monster, is the Camelback kame. By all means hike over to it and take the path that leads up its spine to the windy summit. As you get close, you may notice what intrigued me on my first visit: its hillside has a faint terrace pattern. When not obscured by snow or growing vegetation, there is a definite succession of horizontal levels, as though a giant hand had carefully etched the ridge's flanks with parallel lines. And the view from the top, which reveals much of McHenry County's beguiling bumpiness, is magnificent.

The double-humped Camelback kame at Glacial Park. Approximately 15,500 years ago, the melting Wisconsinan ice front stood just behind the kame.

You can spot three much smaller and somewhat more conical kames fairly close at hand. One stands just southwest of the Camelback; the other two are to the north, in the otherwise flat valley bottom that is part of the park's newly acquired central portion. When you descend the Camelback's southern side, you will also see a relatively small borrow pit excavated before the land was placed in the public trust. Other kames in the region have not been so fortunate; some have been more or less obliterated by quarrying for construction materials.

Kames are one of the classic Pleistocene landforms that owe their origin to the waning days of a has-been glacier. When an ice sheet reaches its maximum extent and begins to stagnate, sand and gravel accumulate near its melting edge. Deposits that fill long, winding stream channels in or under the ice form eskers; those that collect in large heaps, either along the glacier's margins or in depressions in the ice, become kames. Geologists believe the Camelback, like its three diminutive counterparts nearby, is a delta kame—one which formed at the glacier's leading edge. In this interpretation, meltwater spilling off this edge dumped the debris it carried in layer after layer. This characteristic process first made me speculate that the subtle terracing effect noted above was an unusual outward manifestation of its internal stratification. But according to Ed Collins, Glacial Park's resident ecologist, the terracing is more likely a product of the grazing habits of generations of dairy cows and beef cattle that once had free access to the Camelback's flanks. Collins notes that these placid beasts tended to stick together, grazing one after another in lazy, horizontal tracks along the hillside. They eventually created a pattern of walkways reminiscent of contour farming.

The Camelback and its companion kames in the park were formed at a time when the edge of the Wisconsinan ice sheet was fronted by a meltwater lake—a longer, broader, and deeper version of the modern, man-made Wonder Lake, a mile to the southwest. The ancient lake, which probably covered at least the smaller kames, must have persisted even when the ice had retreated some distance to the east. At that point, the Nippersink Valley served as a drainage channel for still more meltwater surging southwestward. When the flood reached the ancient lake's still waters, its velocity would have dropped dramatically. Had this not been so, the kames would have been exposed to the main force of the torrent, and would have been obliterated.

If you explore the smaller kames to the north, you'll probably see old excavations or bare spots revealing sand and gravel as the primary constituents. (On one of my visits, Ed Collins and I discovered that a badger had dug a network of tunnel entrances on the side of one of these kames. The

An exposure of Late Wisconsinan glacial till along a county road south of Huntley. Till, a jumble of rock debris, is the unsorted, unlayered material that was deposited under or directly in front of a glacier.

material the badger had obligingly excavated for our inspection was a fine yellow sand that had been beautifully winnowed and sorted by a glacier-top stream before it was deposited here.) The makeup of the Camelback and the connecting high ground northeast of it is another matter. Here the sediments appear unsorted and till-like: both angular and rounded cobbles are mixed into a brownish matrix of much finer particles. This substrate is so stony that a person using a posthole digger would give up in disgust. This grossly different material might suggest that the Camelback and its neighboring prominences are as close to being a mini-moraine as a kame complex—at least according to precise geological definitions. In fact, though, landforms created on the edge of a glacier may be made of widely differing things, a reminder that the ice-sheet margin is a complex environment where conditions vary widely over relatively small amounts of time and space. Whatever term you use in describing the Camelback, its sediments were unquestionably transported and dumped without being subjected to the kind of sustained, stream-borne sorting action that determined the composition of the park's northern kames.

Behind the Camelback, to its south and east, are more textbook examples of glacial landforms, ones that exhibit "negative topographic expression." That means they are hollows rather than hills or ridges. These are a troika of kettles, depressions that are (or have been) excellent examples of three different wetland types found in northern, glaciated terrain: fen, bog, and marsh. The fen, just south of the Camelback, once hosted a precious community of rare plants that prefer soggy, alkaline conditions. The alkalinity, interestingly enough, was largely provided by groundwater percolating up to the surface through "sweet" glacial drift rich in carbonate-rock fragments. The hydrostatic pressure from below was sufficient to form an unusual domed structure; a low upward bulge is still noticeable on the kettle's northern side. The McHenry County Conservation District is working to restore the fen to its original condition.

The second kettle, the site of a more or less undisturbed bog, is situated east of the Camelback. It features an acidic environment dominated by sphagnum moss and leatherleaf shrubs. Farther east lies the third kettle, a marsh, inhabited by cattails, bulrushes, raucous red-winged blackbirds, and a host of other organisms that thrive in this lush habitat. Of these three hollows, the bog and the marsh are easily accessible to hikers. Make sure you save enough time to visit the marsh's observation platform and the bog's short boardwalk loop. Both are located on the eastern side of their respective wetlands.

Kettles form in a curious way. To understand the process, picture yourself here a little more than 15,000 years ago. The Wisconsinan glacier has melted back somewhat to the east, but there are still big chunks of ice, like landlubber icebergs, that lie detached from the main glacier. The terrain around them is mantled in a coat of glacial drift delivered by the meltwater streams flowing west. These ice blocks persist for many years; perhaps the drift not only surrounds them but also thickly covers them, thus providing excellent insulation against warm summer weather. Finally, they melt away, leaving behind large depressions in the drift. If you've visited Thoreau's famous Walden Pond kettle in Massachusetts, you know such features can be fairly large. Kettles usually become self-contained ponds with no surface outlet. In a climate such as this, vegetation rapidly moves in. Depending on the exact conditions, the ponds can turn into any one of the three wetland examples found here. In the long run, these wetlands may fill in entirely with organic matter, until all traces of open water are gone.

If your visit to Glacial Park has whetted your curiosity about how the Ice Age molded our modern terrain, take an additional hour or two to explore other areas in McHenry County. An entire book could be devoted to this county's natural wonders. The best known of these is Volo Bog State Natural Area, located south of Fox Lake. This endlessly fascinating kettle peatland has a boardwalk that takes you to the very heart of its unique floating stand of tamarack trees. And just a few minutes to its southwest, near the bank of the Fox River, is Moraine Hills State Park, an excellent place to study the rolling forms near the Fox Lake moraine's western flank.

Farther west, the less cluttered farm country is an open book for the geologically inquisitive. One of my favorite examples of the undulous, swell-and-swale surface of Wisconsinan moraines can be seen along Route 47, near Huntley. To the north, along Route 173 between Harvard and Hebron, the upland views from the Woodstock moraine are especially good. And if you want to see what is probably the county's most perplexing feature, drive due west from the town of Woodstock on Kishwaukee Valley Road. After about 6 miles, turn north on Menge Road. Almost at once you descend into what appears to be a shallow stream valley running east-west across the road, and across what Illinois geologists Ardith Hansel and W. Hilton Johnson have recently described as this state's oldest rampart of the Wisconsinan ice sheet—the 24,000-year-old Marengo moraine. There's just one small problem: nowadays, there's no river or creek here. Some geologists have surmised that this was a subglacial channel, the abandoned bed of a stream that once flowed under the Wisconsinan glacier, building the moraine. Both the stream and its encasing ice sheet have long since departed. Whether that ice sheet will have future counterparts descending from Canada and scraping away these landforms to create their own, no one can say. But more than a few experts think our glacial age has many centuries to run. If so, this rich and rolling land will undergo a major facelift once again.

The slight depression that crosses Menge Road west of Woodstock is the bed of a long-vanished stream that flowed under a stagnant glacier. The high ground in the distance is one of the more elevated portions of the Marengo moraine.

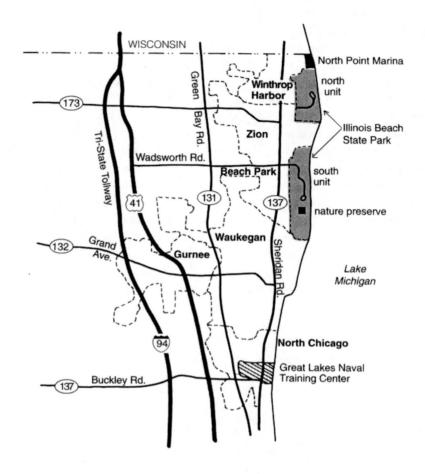

To reach the south unit of Illinois Beach State Park from Sheridan Road in the Beach Park–Zion area, turn east at the Wadsworth Road intersection, pass over the Metra railroad tracks, and follow the roadway as it swings right (south) past the public beach. A sign will guide you to the nature preserve parking lot.

— 9 —
A Pocket History of a Freshwater Sea
ILLINOIS BEACH STATE PARK
Lake County

The wind of the lake shore waits and wanders.
The heave of the shore wind hunches the sand piles.
—Carl Sandburg, "The Windy City"

The advantages of growing up near the Lake Michigan shore are many, but there is also one serious drawback. Later in life, it is difficult for a native of the Illinois lakeshore to move to a region lacking a large body of water. A longtime resident of Wilmette, Rogers Park, or Zion, suddenly transferred to a more landlocked locale, may feel like the proverbial fish out of water. The great inland sea, and the breezes it generates, imprint the souls of those who live at its edge. The lake has also had an important effect on the landforms around it. The shoreline is a remarkably complex and living thing, comprising a multitude of forms, natural and man-made, that are always liable to swift and substantial change.

No place along this coast is more biologically precious, or more illustrative of important geological principles, than Illinois Beach State Park. What a heartening place this is—sunbathers, kite flyers, boaters, bird-watchers, dog walkers, wildflower lovers, ecologists, fishermen—everyone finds something in this sunlit, windswept place to enrich their lives. This essay focuses on what must be holy ground for anyone interested in the earth and life sciences: the nature preserve located in the lower part of the park's southern section.

A network of well-maintained trails leads the hiker through four ecological zones: the modern beach ridge and foredune area skirting the water's edge, the backdune sand prairie, the black-oak savanna established on old beach ridges, and the wet prairie and marsh fringing the Dead River. A distinct pattern characterizes the wooded savanna zone. The pattern consists of a series of low, parallel ridges that arc from north to southwest. These ridges mark the lake's earlier shorelines.

The best place to visualize the geologic forces at work on this coast is the wooden observation platform located near the far end of the foredunes trail, a little north of the Dead River's final approach to Lake Michigan. As you survey the scene from this panoramic perch, pay special attention

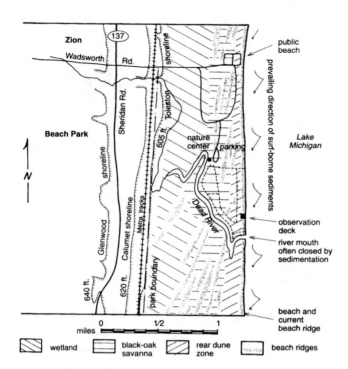

Approaching Illinois Beach State Park, you can see three former lake shorelines. The validity of the Toleston, the lowest shoreline at 605 feet, has been questioned by some geologists.

Map labels: Zion; 137; Wadsworth Rd.; Beach Park; N; Sheridan Rd.; Toleston; 605 ft.; Glenwood shoreline; Calumet shoreline; Metra tracks; 640 ft.; 620 ft.; Park boundary; nature center; parking; Dead River; shoreline; prevailing direction of surf-borne sediments; public beach; Lake Michigan; observation deck; river mouth often closed by sedimentation; beach and current beach ridge; 0 1/2 1 miles

Legend: wetland; black-oak savanna; rear dune zone; beach ridges

to the direction of the longshore currents, as revealed by the direction of the incoming waves. Does the surf meet the beach directly head-on, or does it come in at an angle? More often than not, the latter is the case. For a moment, visualize yourself not as a detached observer but rather as a participant in the lake's great energy machine. You are a single sand particle borne along in the rush of blue water: you sense you are being pushed toward the beach at a southwestward slant. Before long, the wave carrying you drags along the shallow bottom and hurls you onto the hissing beach. But your visit on land lasts only an instant; the backwash pulls you lakeward again, this time in a southeastward direction. The next line of waves shoves you shoreward and southwestward again. And it goes on and on, back and forth.

The net effect is that you are transported a good distance down the coast, even if in a roundabout, zigzag fashion. Eventually, however, you come to rest on the shore. At this point an even more playful manipulator of sand, the wind, picks you up and flings you over the crest of the nearest foredune. You alight at the base of the dune's slipface—the steep backslope. Soon other newly arrived sand grains blanket you; together you form a layer that is part of the dune's characteristic, internal crossbedding pattern. At long

68

On the landward side of foredunes at the nature preserve. The unvegetated area at left center is a small blowout excavated by the wind.

last, you think you've found a stable existence; but in this environment sudden change is always possible. Sooner or later, a particularly violent storm or the rising lake will reclaim you, and once again you'll be suspended in the water's domain.

In general terms, there are two types of coastlines—those where erosion is paramount, and those where the accretion of sediments is actively building up a beach. To geologists, Illinois Beach State Park is particularly fascinating because it features both of these regimes. From the park's northern boundary to a point roughly in line with Wadsworth Road, the shore is, in an overall sense, eroding. Closer to the Wisconsin line, the lake has attacked the land with the voraciousness of a crocodile coming off a vegetarian diet. In 1989, particularly dramatic erosion in one stretch near Winthrop Harbor's North Point Marina showed just what the lake can do when it tries. In the span of a mere nine months, the preexisting shoreline was chewed away a staggering 65 yards westward. (In the winter of 1995–1996 I visited this site repeatedly, to watch a seemingly endless string of dump trucks lay load after load of sand on the beach, in an effort to counteract the lake's all-too-powerful assault.)

South of Wadsworth Road, though, more waterborne sediments arrive than depart. A hundred years ago, the observation deck would have been much closer to the water's edge, or perhaps even past it. Experts predict that

Pebbles on a Lake Michigan beach. The bright white pebble are Silurian Niagaran series dolomite and the dark stones are igneous rocks—diabase or basalt. Other pebble varieties include chert, granite, and translucent quartz. The prominent specimen at center, which might be mistaken for a sedimentary rock known as conglomerate, is a smoothed lump of man-made aggregate.

the building-up of the shore in this stretch will eventually give way to net erosion in the first few decades of the twenty-first century. If you're reading this book in A.D. 2030, you might check to see whether they were right. Regardless of which part of the coast is gaining or losing ground, you have to wonder where all the sand, gravel, and pebbles come from in the first place.

A stroll over the modern beach ridge and along the water's edge reveals an important clue in the form of pebbles that lie stranded in the finer sediment. These lovely objects, streamlined and rounded by the ceaseless action of the surf, together constitute an open-air geological museum. People adept at reading the stories they contain have a vista of more than two billion years before them. Next to examples of chaste white Silurian dolomite—perhaps originally from the Niagaran Escarpment of Wisconsin's Door County—lie two other types that might pass for pockmarked pumice and red sandstone. They are, respectively, lumps of steel-mill slag and well-worn pieces of old brick: johnny-come-lately human additions to this display of much older natural materials. But also present are the ancient ones: igneous and metamorphic rocks of Precambrian age that hearken back to distant episodes of seafloor spreading, continental collision, and mountain building. Much more recently, they have ridden the Pleistocene ice sheet from the distant reaches of the Canadian Shield. It is these handsome specimens—pink granite, quartzite, greenstone, and basalt and diabase black as night—that offer the best proof that glaciers brought huge quantities of rock debris far southward, down through the Lake Michigan basin. If you scrutinize many of the beach stones, you may find more obvious evidence of the glacial mode of transport. A few of the pebbles bear deep, linear scratches called glacial striations. These grooves were formed when the stones became lodged at the base of the slowly moving ice, where they were ground against other rocky surfaces.

No one can say how many times the Pleistocene glaciers came past this point. The record reveals only selected details of that epoch's last glaciation, the Wisconsinan. The ice sheet advanced into northeastern Illinois some 25,000 years ago. From roughly 16,000 to 13,500 years ago, the Wisconsinan ice sheet staged its final advances and retreats in this region, creating the Valparaiso, Tinley, and Lake Border end moraines that are still evident west of here. Beginning about 14,000 years ago and extending to about 12,500 years ago, Lake Chicago, Lake Michigan's predecessor, apparently reached its greatest extent on two separate occasions, in the so-called Glenwood I and Glenwood II phases. At these times, the lake surface stood at a lofty 640 feet above mean sea level—about 60 feet higher than it is today. Much of the Lake Michigan coast as we know it was flooded. Almost all of the land now making up the city of Chicago was under the waves. During the Glenwood phases, the swelling meltwaters from the Lake Michigan ice lobe were probably supplemented by runoff from the more eastern Saginaw and Erie lobes.

Other significant changes were in the offing. By 12,200 years ago, the ice had retreated north of Michigan's Lower Peninsula, and the lake could drain eastward toward the North Atlantic. Four hundred years after that, the ice once again blocked this northern outlet; this time Lake Chicago rose to about 620 feet—the Calumet phase. In the millennia that followed, the lake's shimmering surface sank and rose again and again, but it never wholly returned to its impressive Glenwood or Calumet levels. Still, about 4,500 years ago, during the Lake Nipissing phase, the water rose high enough to sweep away much of the till and other sediments that the glacier had previously dumped near the shore. It is largely this material, sorted by wave action and to some extent by the wind, that now forms modern Lake Michigan's beaches.

As you might expect, this record of ever-changing lake levels and wavering ice fronts has proven difficult to decipher. Whatever the woes in your life, take a few seconds to give thanks that you are not a Quaternary geologist. Over the years, chronologies and interpretations of Pleistocene and Holocene events, based on stratigraphic analysis, radiocarbon dating, and other methods, have been almost as numerous as the stones on this beach; and like the stones, no two are exactly alike. Another complicating factor, known as isostatic rebound, occurs when a glacier reaches its full extent and its weight actually pushes the crust down into the underlying, yielding mantle. In this region, when the ice retreated, the burden on the crust was removed, and it gradually began to rise again. To what extent has the slow lifting of the land affected the lake's level, drainage patterns, and erosional effects? For now, we can't be sure.

Before you leave the nature preserve, take a few minutes to cautiously explore the lower reaches of the well-named Dead River. Most of the time, this waterway suffers from low self-esteem and cannot even rouse itself to reach the lake. Its mouth is blocked by a sandspit constructed by the south-ward march of sediments. Once in a while, however, high water levels or storms make this a busy place. In the spring of 1995, I saw the Dead River pouring through the breached spit. A temporary drainage channel in a trough along the beach fed the river mouth with pulse after pulse of rain and surf water that carried a considerable quantity of suspended fine sediments. These sediments came to rest in a broad, fanlike delta on the north side of the river mouth. When I returned the following day, the sun was shining, the lake was calm, and the delta looked for all the world like terra firma. But when I stepped out onto the new landform's landward side, I sank to almost hip level before I managed to clamber out of it. May that prove to be my one experience with quicksand—material that appears stable at the surface, but is supersaturated with water.

On your return leg through the backdune zone, see if you can find the unvegetated hollows (they're called blowouts) where the wind is still actively excavating and changing the land's surface. Return to the parking lot through the pleasant, leaf-dappled black-oak savanna and concentrate on the old beach ridges, the sand dunes that sit atop them, and the interdunal ponds that lie in the depressions between them.

View facing westward toward the backdunes and the black-oak savanna, in the park's nature preserve. Former Lake Michigan beach ridges are most apparent in the savanna zone.

This vertical sandy bank of the Dead River was located where the stream cut through a beach bar that normally blocks access to the lake. The many bedding layers suggest the shore is being built up with fresh sand. The zone of gravel likely represents a former pebble-strewn beach surface similar to the current one, visible at the top of the photo. Note the faint ripple-mark pattern on the present-day surface.

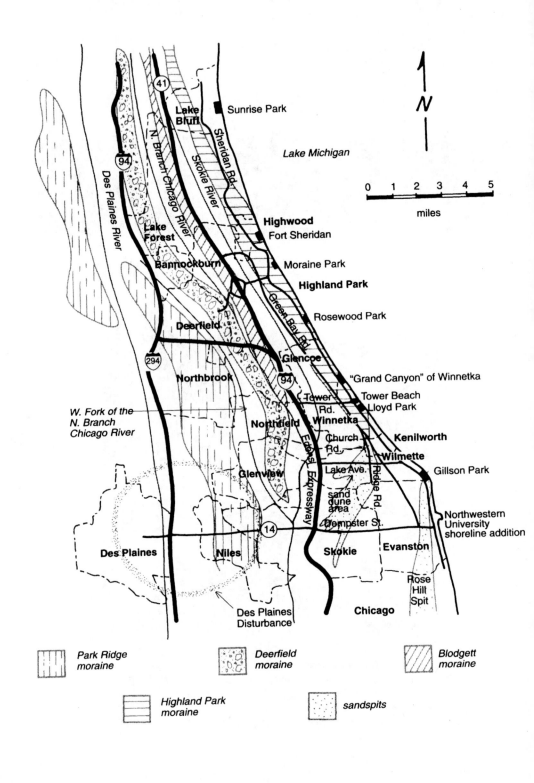

Map Labels

(41)
(94)

Sunrise Park

Lake Bluff

Des Plaines River

N. Branch Chicago River

Sheridan Rd.

Skokie River

Lake Michigan

0 1 2 3 4 5
miles

Highwood
Fort Sheridan

Lake Forest

Bannockburn

Moraine Park

Highland Park

Green Bay Rd.

Deerfield

Rosewood Park

(294)

Glencoe

(94)

Northbrook

"Grand Canyon" of Winnetka

Tower Rd.

Tower Beach
Lloyd Park

W. Fork of the N. Branch Chicago River

Northfield

Winnetka

Kenilworth

Church Rd.

Wilmette

Edens Expressway

Lake Ave.

Ridge Rd.

Gillson Park

Glenview

sand dune area

(14)

Dempster St.

Northwestern University shoreline addition

Des Plaines

Niles

Skokie

Evanston

Rose Hill Spit

Des Plaines Disturbance

Chicago

Legend

▦ Park Ridge moraine

⠿ Deerfield moraine

▨ Blodgett moraine

▤ Highland Park moraine

▦ sandspits

— 10 —
A Primer of North Shore Geology
LAKE BLUFF TO EVANSTON
Lake and Cook Counties

*Blue water wind in summer, come off the blue miles
of lake, carry your inland sea blue fingers,
carry us cool, carry your blue to our homes.*
—Carl Sandburg, "The Windy City"

The communities along the Lake Michigan coast just north of Sandburg's Windy City have a collective identity, the North Shore. This is a fairly compact area of affluent suburbs where independent-thinking and civic-minded townspeople have built one of the nation's most acclaimed public-education and library systems, at no small expense to themselves. The North Shore is a successful amalgam of lovely residential neighborhoods, parks, forest preserves, and unblighted downtown shopping districts. To native and visitor alike, the area's defining element seems to be the ubiquitous greenery that sets the North Shore apart from the road rash and condo clonescapes surrounding it. The North Shore is defined by something else, too: a geographical uniqueness born of the Ice Age. Its juxtaposition of high and low ground, so different from the chessboard flatness of Chicago's lacustrine plain, has played a controlling part in the history of this prestigious area.

Even lifelong residents rarely realize how many ancient landforms here have survived more than a century's worth of intense settlement. These features date from the final phase of the Pleistocene epoch and from the early part of the succeeding Holocene, the current epoch that contains the very short span of human ascendancy. To see some of these features, spend a few hours in a gradual southward descent through the North Shore, beginning at Lake Bluff's Sunrise Park. This lovely parcel of public land is located on the Lake Michigan shoreline at the foot of Center Avenue. If you have already been to the more northern Illinois Beach State Park, the site described in the preceding essay, you will note at once a distinct difference. Instead of standing on a low, broad, dune-dotted shore, you are surveying the lake's blue surface from a high bluff. Illinois Beach is situated in what geographers call the Zion Beach Ridge Plain, but this locale is part of the Lake Border Moraines Bluff Coast. The Bluff Coast extends from

Groins placed along the Lake Michigan shore at Lake Bluff's Sunrise Park. There is more sand and gravel on the southern side of each groin than on the northern side.

North Chicago to Winnetka, where it gives way to the Chicago-Calumet Lacustrine Plain.

Sunrise Park has a beach of sorts, but it only dates from the 1970s, when the unusually high lake level and increased shoreline erosion prompted the building of a new barrier to the south. This structure was installed to protect the unconsolidated material (actually three different units of Wisconsinan till) that holds up the bluff face. Such man-made barriers, including long harbor piers and the short groins you see sticking out from this beach, trap sediments that would otherwise migrate southward along the shore. Today, this narrow cushion of sand and pebbles is Sunrise Park's main way of preventing Lake Michigan from gobbling up its scenic heights. The till that was once open to view along the bluff face is largely grassed over—the result of civic vegetation projects and our wet continental climate. Unfortunately for geologists, the vegetation obliterates any patch of freshly exposed rock or sediment that would interest them.

When you're ready to push on, depart Sunrise Park by proceeding due west on Center Avenue, through a particularly handsome neighborhood. When you reach Lake Bluff's small downtown area, which looks as if it was airlifted lock, stock, and barrel from rural New England, bear right by the city hall until you face the Metra commuter-train station at the intersection of Sheridan Road. (Avoid taking the Rockland Road/Route 176 underpass

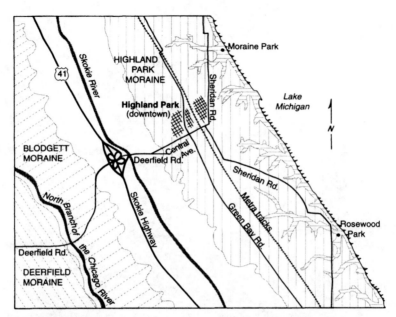

Geologic map of the Highland Park area. Note the extent of the ravines on the eastern side of the Highland Park moraine. If left to their own devices, these steeply pitched streams would eventually break through the moraine and capture the Skokie River, causing the latter to flow directly into Lake Michigan. Also note the many small projections along the lakeshore. Each represents a separate pier or groin constructed to prevent shoreline erosion.

under the tracks.) Turn left (south) on Sheridan Road and follow its often devious wanderings attentively. After passing through the Lake Forest College campus and superb ravine country dotted with palatial homes, you will enter Highwood and skirt the now largely inactive Fort Sheridan army base.

In downtown Highwood, Sheridan Road takes another jog lakeward; watch carefully for the signs. About 0.3 mile after the road resumes its southward direction at Fort Sheridan's lower boundary, you'll cross a large bridge over a major ravine. Turn left immediately into Moraine Park, one of many sites in the city of Highland Park's superb and extensive system of public open spaces. Here you can explore the ravine's final descent to the lake. To the west, in the valley of the Skokie River, the land is extremely flat; the river drops as little as 2 feet per mile. In ravines such as this, however, the gradient is up to fifty times as steep—100 feet per mile.

Ravines are among the Bluff Coast's most characteristic features. These precipitous gullies have been carved by youthful streams that are doing their

The North Shore coast at Winnetka's Tower Beach on a glum winter's day. The tree-covered Highland Park moraine—already partly removed by lake erosion—marches up the shore, into the misty distance. The dark "boulders" in the foreground are actually detached blocks of shelf ice encrusted with sand and small pebbles.

best to cut through the high ground, in response to the low base level established by the modern lake elevation. The gullies—there are more than thirty of them along the Illinois lakeshore—have a more or less V-shaped cross-sectional profile. While most gullies can be glimpsed only from the bridges that span them, there are two other sites off the main track of this excursion that are well worth side trips. The first is the secluded Rosewood Park, on the east side of Sheridan Road about 1 mile north of the Highland Park–Glencoe line. Here you can take a wooded path up another impressive ravine, from almost lake level to the blufftop. In spring you can see the unique ravine wildflower community that in so many other places has been extirpated. The second location, farther south, is the section of Sheridan Road just south of Glencoe. I call it the Grand Canyon of Winnetka. The main roadway plunges through a forking ravine in a way that delights both the adult geologist and the ten-year-old daredevil bicyclist. Take it from someone who has been both.

When you're done at Moraine Park, resume your southward course on Sheridan Road to Highland Park's bustling business district. After turning right onto Central Avenue, cross the Metra tracks and continue west on Central. In the vicinity of the intersection with Green Bay Road, you pass over the crest of the Highland Park moraine, the easternmost continuous component of the Lake Border morainic system. From Lake Forest to Winnetka, the Highland Park moraine extends from west of Green Bay Road eastward to the lakeshore. In fact, lake erosion has chewed away a portion of the ridge's backslope. At this spot, in downtown Highland Park, the moraine shows some of its most dramatic relief; you are now about 120 feet above Lake Michigan.

The moraine has played a major role in the region's development. Green Bay Road, which hugs its spine for several miles north and south of here, is one of the North Shore's high points in a historic sense as well. In the early days of the republic, the thoroughfare linked Forts Dearborn and Howard—now the homes of the Bears and the Packers, respectively.

After the Green Bay junction, Central Avenue quickly descends the moraine's steep foreslope. Soon you'll merge onto Deerfield Road, and soon after that you'll reach the cloverleaf intersection of Route 41 (Skokie Highway). Keep right and take the ramp onto Route 41 South. You are now in the valley of the Skokie River, the lowland between the Highland Park and Blodgett moraines. At the highest (Glenwood) phases of ancient Lake Chicago, this valley was a narrow arm of the lake, and a great deal of organic debris, which geologists classify as Grayslake peat, accumulated on its bottom.

After the lake's retreat, this zone was prime wetland. Before human population pressure intensified in the late twentieth century, the area was largely given over to a strange-bedfellows mixture of forest preserve, transportation corridor, and industrial use. This arrangement reflects the fact that in the North Shore's formative period, homeowners and farmers considered the high ground atop the flanking moraines not only more scenic and more easily tilled but also healthier. The low wetland was regarded (with some justification) as unwholesome, prone to flooding, and pestilential. A string of splendid nature preserves was created by the Depression-era Civilian Conservation Corps without much protest either from business or landowning interests. Nowadays, most North Shore residents would say that the somnolent beauty of the Chicago Botanic Garden complex and the Skokie Lagoons forest preserve is one of the most wholesome aspects of the region. Most urban and suburban people see wetlands not as a dangerous and unproductive manifestation of hostile nature but as a priceless natural and recreational resource. As the Skokie Valley becomes more and more developed, you have to wonder how much of this new construction would have been attempted were it not for the invention of the automatic sump pump, which keeps the high wetland water table from turning basements into indoor swimming pools.

The Blodgett moraine, which here lies just west of Route 41, is by no means the westernmost member of the Lake Border complex. Beyond it lies the valley of the North Branch of the Chicago River; beyond that are the Deerfield and Park Ridge moraines and the valley of the Des Plaines River. Still farther west, beyond the Lake Border complex, lie the Tinley moraine and the lofty Valparaiso morainic system, which circles the Lake Michigan basin like a giant bathtub ring. When I show visitors the Lake Border moraine area and its relatively subtle landforms, I have trouble persuading them that until recently this area contained a continental divide as significant as the one in the mighty Rocky Mountains. Since the beginning of the twentieth century, when the Chicago Sanitary and Ship Canal was opened, the flow from the North Branch and the Skokie River has been diverted to the canal and the lower Des Plaines River, and in the end to the Illinois and Mississippi Rivers. But before that unnatural act of engineering was completed, these two local streams fed Lake Michigan.

To understand how the divide worked for thousands of years, visualize a thunderstorm passing over this area. One moment, raindrops would fall west of the Park Ridge and Deerfield moraines. A moment later, more raindrops would fall to the moraines' east. The first, western set of raindrops would have made their way down the Des Plaines, Illinois, and Mississippi Rivers to the Gulf of Mexico; while the second, eastern set would have proceeded via the Skokie River and the North Branch to Lake Michigan, the lower Great Lakes, the St. Lawrence River, and the North Atlantic. That's a big difference in destinies.

To the geologist, the southern portion of the North Shore is as fascinating as the northern. Route 41 becomes Edens Expressway where it joins Interstate 94 south of Lake-Cook Road. Remain on the highway until you reach the Tower Road exit; then take Tower east. You will soon pass through the Skokie Lagoon complex, where a dam has backed up the Skokie River to produce a humble simulation of Lake Chicago's incursion. This peat-floored preserve is a favorite spot for local bicyclists and canoeists.

Continue east on Tower Road through Winnetka's Hubbard Woods district, and to the road's terminus at Sheridan Road. There, at Tower Beach, you can examine the truncated rear side of the Highland Park moraine, which forms a high rampart northward along the coast. Just a little bit farther down Sheridan Road lies the entrance to Lloyd Park, where the moraine

After fronting the Lake Michigan coast for most of the length of the upper North Shore, the Highland Park moraine cuts inland and due south at Lloyd Park. One can trace its wave-cut foreslope—seen here as a gentle rise in the background—through much of southern Winnetka.

finally quits its lakefront station and angles due south and inland. If you follow the eastern flank of the ridge through southern Winnetka, you will note that it forms a relatively steep slope. This is a wave-cut cliff that was shaped when the surf of Lake Chicago was at its highest, Glenwood stages, from about 14,500 to 12,200 years ago. The cliff extends across Green Bay Road, through the North Shore Country Day School campus, and up Church Road to the vicinity of the Indian Hill Country Club. This is the southern tip of the Highland Park moraine. If you continue down on Ridge Road (the continuation of Church Road), you will still be on high ground, moving along the crest of the Wilmette spit, a long, curving body of sand and gravel stretching southwestward all the way to the southern part of Skokie.

A few blocks farther south, you will reach the intersection of Ridge Road, Kenilworth Avenue, and Park Drive. This is a good place to spend a moment pondering the landform under your feet. As you face eastward and look down Park Drive, you are beholding the old floor of Lake Chicago. Like Winnetka's wave-cut cliff, the Wilmette spit is a product of the maximum high water of the Glenwood phases, when Lake Chicago reached 640 feet above sea level—a whopping 60 feet above modern Lake Michigan. The spit was created when longshore currents swept around the tip of the moraine and deposited their sediments beyond it. When the lake receded, the spit was exposed. During Lake Chicago's later Calumet phase (about 11,800 years ago), the lake rose back to 620 feet, so this ridgetop was then a narrow jetty of land.

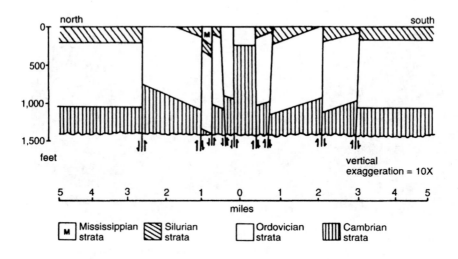

Simplified cross section of the earth's crust in and around the Des Plaines Disturbance. This pattern of severe local deformation is characteristic of a meteorite impact site. —Illinois State Geological Survey

In one spot, sand dunes formed atop the Wilmette spit. To see the still somewhat hummocky surface that marks their presence, head farther south on Ridge Road, into the old Gross Point section of Wilmette. The dune field begins near the intersection of Wilmette Avenue and Illinois Road and extends onto the grounds of the Westmoreland Country Club. This blanket of dry, easily excavated sand has posed no obstacles whatsoever to residential construction.

Now backtrack lakeward on Wilmette Avenue. Just after you cross Ridge Road, you'll descend from the spit onto the Chicago-Calumet Lacustrine Plain. Continue through Wilmette's downtown area until you reach Sheridan Road, and turn onto it heading south. Sheridan takes you past Wilmette's ever-popular Gillson Park. Next, you wind around Wilmette's famous landmark, the Baha'i Temple, where you briefly rise onto the spine of a later, lower counterpart of the Wilmette spit, the Rose Hill spit, and enter Evanston, a handsome satellite city that has stoutly resisted annexation by its giant neighbor to the south.

As you proceed farther south on Sheridan, passing through the Northwestern University campus, you may agree that few institutions of higher learning enjoy a more handsome site. Sadly, this ideal waterfront setting includes one example of humanity's tampering with the lakeshore's land forms and ecosystems. When the university sought to expand its campus acreage in 1962, it acquired a large amount of sand quarried from the magnificent dune fields east of Gary, Indiana. The campus expansion project was hotly opposed by one of Illinois's staunch environmental advocates, Senator Paul Douglas, as well as university staff members, alumni, and students, who protested that the landfill would damage one of the Midwest's most beautiful and fragile habitats. Nevertheless, the project went ahead. On a topographic map, the extension presents a distinctly incongruous profile.

A significant (but invisible) geologic feature, the Des Plaines Disturbance, lies about 10 miles west of the Northwestern campus. This is a zone where the earth's crust has sustained intense faulting and deformation in a relatively small area. Centered near the intersection of Dempster Street and the Tri-State Tollway (Interstate 294), this strange structure is approximately 5.5 miles in diameter. The strata have been subjected to some force that caused a remarkable degree of fracturing and up-and-down displacement between the fault blocks. Over the decades, earth scientists have offered several theories to account for this most unusual feature. Recent evidence, in the form of well-core samples containing rock fragments with shatter-cone patterns, points to a major meteorite impact that occurred at some point after the Pennsylvanian, the last period of large-scale rock formation in northeastern Illinois.

The presence of the Des Plaines Disturbance has been revealed only by a painstaking analysis of well logs. The disturbance is not something you will be able to spot at the surface. In the long span of time since the meteorite crashed to earth, the crater and the uppermost rock strata have been eroded to the same level as the surrounding terrain. This odd subsurface feature—together with similar impact structures in Glasford, southwest of Peoria, and Kentland, Indiana, about 55 miles south of Gary—is a powerful example of how the earth has been affected by extraterrestrial bombardment over the course of geologic time.

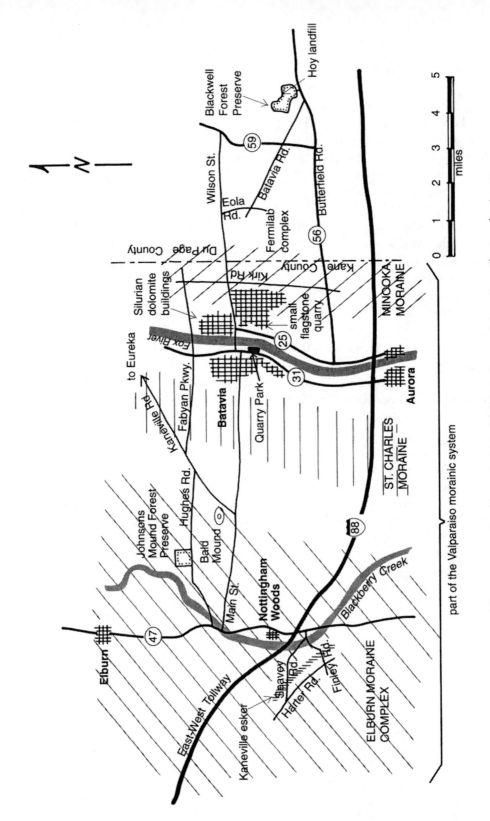

The middle Fox River valley, situated within the Valparaiso morainic system. The tour begins at Johnsons Mound Forest Preserve on Hughes Road east of Illinois 47.

part of the Valparaiso morainic system

— 11 —
In the Land of the Fox
BATAVIA AND ENVIRONS
Kane and Du Page Counties

It is certain that . . . skeletons of elephants have been found, and it is
probable that more will be found in Canada, in the country of the
Illinois, in Mexico, and in other places in North America.
> —George-Louis Leclerc, Comte de Buffon,
> *Époques de la nature,* 1807

The geologic history of the middle Fox Valley is as abundantly fertile as the farmland that flanks its beautiful river. Each of the commuter towns and satellite cities that adorn Chicagoland's westernmost waterway have excellent quarries, moraines, and river-formed features that together tell a story with two large chapters: the early Paleozoic era, when the region's hard bedrock formed at the bottom of a tropical sea, and the glacier-ridden Pleistocene epoch. In keeping with this book's modest size and mission, I present only a fraction of all the sights to explore between Carpentersville and Aurora. Consider this an hors d'oeuvre designed to whet your appetite. For the main course, turn to the more detailed Illinois State Geological Survey field-trip guides to Batavia, Dundee, Elgin, Elmhurst-Naperville, and neighboring communities. Other publications and maps (including U.S. Geological Survey topographic maps) will help you unlock clues to the Fox Valley's past. And if you get the opportunity to take one of the free public field trips offered in this area by the Educational Extension of the State Geological Survey, by all means do so.

The comely riverfront town of Batavia is the essay's geographical center point. However, we begin not in its valley setting but a little to the west, on the broad upland of Blackberry Township, at Kane County's Johnsons Mound Forest Preserve. This small parcel of public land, on Hughes Road a couple of miles east of Route 47, consists of a single heavily wooded hill. When you reach it, speculate on how this prominence—which stands atop the Wisconsinan till of the Elburn moraine—came to be. One clue to its origin can be found a mile to the southeast, where heavy quarrying reveals the sand and gravel deposits of its treeless counterpart, Bald Mound. To the geologist, the nature of the sediments exposed there is a big tip-off: it suggests

A farm spread sits atop one section of the Kaneville esker, the longest landform of its type in the state.

that Bald Mound, and probably Johnsons Mound as well, are kames—intriguing Ice Age landforms created near the margin of a glacier that had stopped advancing. If you've visited McHenry County's Glacial Park (essay 8), you've already seen one kind of kame: the delta type, formed by meltwater spilling over the front edge of the ice sheet to form a ridgelike feature. Johnsons Mound is circular or oval rather than ridgelike, so it likely formed when meltwater-borne sediments accumulated in a separate hole or depression in the body of the glacier.

Kames are sometimes found in association with other important stagnant-ice landforms—the low, long, and linear eskers. They, too, are composed of the clean, well-sorted sand and gravel essential in construction. When not preserved in the public trust, as at Johnsons Mound, these landforms soon fall before the excavator's mechanized blade and shovel. Were Congress ever to enact an Endangered Landforms Act, kames and eskers would be on the short list.

The origin of eskers has been the subject of several theories over the years. Most glacial geomorphologists believe that they represent sinuous stream channels in or below the glacier that eventually became plugged up with sediments. The longest esker in Illinois is close at hand. If you take Route 47 south just past Interstate 88 and turn west on unpaved Finley Road, you will soon cross Blackberry Creek and find yourself staring at the renowned Kaneville esker. Farmers often built their spreads on high ground,

Tufa, the dark, spongy precipitate on the Silurian dolomite, is visible along the western face of Batavia's Quarry Park.

like the farm directly ahead on the crest. This elongated ridge runs northwest and southeast. It is not protected from mining, and a county highway department facility is located where Seavey Road slices through it.

When you return to Batavia, stay on its western bank. At the base of Union Street, just off Route 31, you'll find the entrance to Quarry Park, a riverfront recreational facility that sits on the floor of a long-abandoned pit cut into Silurian dolomite of the Sugar Run formation. Along the park's western wall you'll find a long, weathered outcrop of this roughly 420-million-year-old rock.

Dolomite is a close relative of limestone that contains more magnesium carbonate than calcium carbonate. From the look of things here, there is enough calcium carbonate to cause the precipitation of tufa, a spongy deposit that here is dark brown to black. You should spot it easily. Tufa forms in several ways. In this case, water has seeped through the dolomite, dissolved some of its carbonate content, and then deposited it on the rock face. One dense form of tufa, called travertine, has eye-catching banded patterns and is highly prized as polished ornamental stone.

Quarry Park's stone-producing days are long over, but Sugar Run dolomite is still being extracted, at least sporadically, across the river on Route 25 a few blocks south of the Batavia bridge. Here the rock reveals its thin-bedded nature, which makes it popular as flagstone. If you look

A small flagstone quarry on the east side of the Fox River, in Batavia. Here, the Silurian Niagaran series dolomite is typically thin-layered and slablike. Note the Pleistocene sediments that lie directly atop the Silurian strata.

closely at the quarry cut face, you'll note that a much younger deposit of unconsolidated glacial drift overlies the bedrock.

The final attraction lies east of Batavia. Before you leave town, though, note that some of the older buildings by the river are made of the distinctive Sugar Run dolomite known rather inaccurately as Joliet marble (if it was quarried in that town) or Athens marble (if it came from quarries in the town of Lemont). This architectural stone weathers to a buttery yellowish tint; if it makes you think of the Chicago Water Tower, there's a good reason (see essay 12). Nowadays, dolomite is not widely used for building cladding, because of its high cost and low availability. Some architects shy away from it because it tends to exfoliate, or peel off in layers, as it ages.

Proceed up the eastern Fox River bluff on Wilson Street. At the Kirk Road junction, you are crossing over the crest of the Minooka moraine. The road continues down the moraine's back slope and past the renowned Fermilab National Accelerator complex. At this site, where only a few thousand years ago a glacier surged forward, stalled, and stayed put long enough to dump load after load of rock debris, human beings now manipulate subatomic particles with the same deftness their ancestors showed in fashioning tools from Paleozoic chert. Here, in these two different spheres, the human and the geological, we see how rapidly the world has changed in this latest chapter of our planet's history.

The grand reward of this trip is Du Page County's Blackwell Forest Preserve, a favorite spot for the residents of the residential communities around it. The preserve is about 3 miles east of Fermilab. Enter it on the south side, off Butterfield Road just east of the intersection with Batavia Road. At once you'll see it is dominated by a solitary hill, Mount Hoy, somewhat reminiscent of Johnsons Mound. Is this, too, a kame, wisely preserved for the geological enlightenment of the public? If you had permission and the right equipment, you might make a test boring to see if this landform is made, as all good kames are, of gravel and sand. But if you did, you'd likely retrieve a sample containing scraps of plastic, glass fragments, and aluminum beer-can tabs. This unique geological deposit dates not from the Pleistocene, but from the mid-1960s to the early 1970s.

In those years, Mount Hoy was a sanitary landfill, also known as the town dump. According to the Illinois State Geological Survey, this impressive hill is (or was) almost one-third paper. Yard and food wastes together account for even more—about two-fifths. Glass, iron, and steel objects take up another fifth. Of all these, glass is the most impressively resistant to change; essentially it remains glass in perpetuity.

Mount Hoy, the kamelike, conical hill across Blackwell Preserve's Silver Lake, is a defunct sanitary landfill that now serves as a recreational site.

Landfills have a distinctly geologic nature, and they present distinctly geological problems. This neat, well-tended hill and preserve give the impression that once the last refuse was buried, there was no further cause for concern. As the refuse in a landfill rots or otherwise breaks down, however, an unsavory liquid by-product, leachate, forms. This authentically disgusting substance is usually dark red; it stinks of impure methane; it often swarms with bacteria and toxic chemicals.

It is imperative to contain leachate on site, so it does not pollute aquifers (water-bearing strata or sediment deposits) or nearby streams. One way to minimize the danger while the landfill is still active is to cover each day's load of trash with its own layer of impermeable clay. This encapsulates each cell of decaying matter in much the same way that a warship's hull is divided into many watertight compartments. Some leachate always seems to escape, anyway. A trench dug around the landfill perimeter can be an effective form of entrapment—and in fact Mount Hoy has such a barrier. State environmental laws also require the periodic collection of water samples off the site, to see if any leachate is migrating toward water supplies despite all these precautions. But while county and state officials here work to keep potential pollution in check, the public enjoys this well-situated piece of high ground. In the winter, the clear and frosty air is filled with the shrieks of kids inner-tubing down Mount Hoy's snowy flanks. Little do they realize that this man-made hill is shrinking very slowly beneath them, as microorganisms, snug in the cozy warmth generated by decay, continue to break down the abandoned products of civilization.

Not all points of geologic interest at Blackwell are so recent in origin. In 1977, while excavating clay fill out of the preserve's McKee Marsh, a worker unearthed the skeleton of one of the heftiest animals of the Pleistocene epoch—a woolly mammoth. The animal's remains now reside at Oak Brook's Fullersburg Woods Environmental Education Center. According to State Geological Survey radiocarbon dating, this beast lived about 13,000 years ago. That was when the Wisconsinan ice sheet had retreated out of northeastern Illinois, and when the high waters of ancient Lake Chicago were periodically breaching moraine dams to the south of here and flooding the Illinois River valley. In examining the plant remains associated with many mammoth-skeleton sites, paleontologists have learned that these great herbivores were primarily tundra dwellers. They were well adapted to life in this exposed environment: besides being well insulated with fat and double-coated fur, they had relatively small ears and feet, which minimized heat loss, and massive grinding teeth for breaking up the tough, low-lying vegetation. In contrast, the other familiar "elephant" that inhabited Pleistocene Illinois, the mastodon, preferred taiga (northern coniferous forest) conditions and dined primarily on black spruce and larch. Specialists now believe that the mastodon was sufficiently different from the mammoth and modern elephants to place it in another family.

Does the presence of a mammoth here in Du Page County mean that the whole area was tundra in prehistory times? Not necessarily. The studies of paleoecologists and pollen experts suggest that the land south of the ice sheet was not an unbroken band of tundra, as was once surmised, but a complex patchwork pattern of tundra and taiga forest. Despite the challenging conditions, this mixture of environments was home to a large number of mammals. But then, in the short period between 12,000 and 9,000 years ago, approximately forty mammalian species vanished from Ice Age North America without a trace. Were they hunted to oblivion by our continent's earliest human inhabitants? Or were they incapable of adapting to the sudden climatic reverses of the Pleistocene and the early Holocene? The decades-long debate inspired by these questions has been heated, and still rages.

The Architectural Geology of Downtown Chicago: A Sampler
THE NEAR NORTH SIDE AND THE LOOP
Cook County

Not to find one's way in a city may well be uninteresting and banal.
It requires ignorance—nothing more. But to lose oneself in a city—as
one loses oneself in a forest—that calls for quite a different schooling.
—Walter Benjamin, "A Berlin Chronicle"

Earth science is an octopuslike subject, with its tentacles reaching into many unexpected places. To most people, geologic exploration implies a wild setting: the layer-cake chasms of Arizona's Grand Canyon, the volcanoes of Iceland, the glacier bays of Alaska. Few places, though, are as geologically rewarding as the heart of a great city, where the forces of economics, political power, and civic pride combine in the uniquely geological art form of architecture. For the person interested in the scientific underpinnings of architecture, losing oneself in downtown Chicago is much like losing oneself in paradise.

When we think of the primary materials that shape our modern civilization, we think mainly of steel, plastic, and the microchip. But no other culture has relied on stone products as comprehensively and as voraciously as ours. If we used our terms more precisely, we would call this the Stone Age. According to the State Geological Survey, in one average year in the recent past, Prairie State quarries produced the equivalent of more than five tons of stone for each Illinois resident. These quarries supply massive amounts of sedimentary rock—Ordovician sandstone, Silurian dolomite, and Mississippian limestone—for a wide array of applications, from massive blocks destined for Lake Michigan piers to fine quartz grains used in glassmaking. In Chicago's monumental heart, though, the ornamental stone comes most often from sources outside the region, and even outside the nation. Here in the urban core, one can get an excellent glimpse of the geology of the greater world.

In contrast to the ridged morainal country to the north and west, Chicago sits on the almost perfectly flat floor of ancestral Lakes Chicago and Nipissing—predecessors of Lake Michigan from approximately 14,000 to about 4,000 years ago. When these waters covered the metropolitan area, they deposited a deep layer of fine sediments. As the owners of Chicago's earliest tall buildings soon learned, these silts and clays hardly make an ideal substrate. Anchoring a skyscraper in Chicago is much more difficult than in New York City, where surface exposures of the tough, Paleozoic schist have facilitated the construction of Manhattan's two high-rise districts, Midtown and the Wall Street area. The main danger posed by the Windy City's unconsolidated lakebed deposits is that they tend to settle unevenly when subjected to great weight. No one sitting in a sixty-fifth-floor office wants that sort of uncertainty.

The eleven buildings described in this essay are located within what most readers would regard as a healthy ninety-minute walk. The starting point, the Chicago Water Tower, is Chicago's most celebrated landmark and stands at the focal point of the busy Near North Side. This land just above the final stretch of the Chicago River is a perfect example of the impact human activity has had on the Lake Michigan coastline. Originally, the eastern portion of the Near North Side was either submerged ground or an unappealing patch of coastal sand dunes, depending on the exact locale. In the 1820s, Michigan Avenue stood only a city block or two from the waters of the lake. This locale did not attract the general attention of builders and investors until the 1880s, when one of the most picturesque of the city's many scoundrels, George "Cap" Streeter, grounded his ship on a shoal, roughly where the mighty John Hancock Center now stands. When he realized that he was stranded for good, Streeter convinced local contractors to dump their construction waste around his vessel. Soon the whole area was filled in; and the real estate value of the parcel, known as Streeterville, soon became apparent. Because the land had been snatched from the lake, it was not in Illinois's jurisdiction, or so Streeter contended. He proclaimed his domain an autonomous district and set himself up as its federal governor—an impressive promotion for one who formerly had been an itinerant showman and circus producer. Amazingly—given Chicago's penchant for strong-arm politics—it took three decades for the city's judges and nabobs to dislodge him from his waterfront roost. When he was booted out, he was convicted not of impersonating a U.S. government official or of deluding investors but of selling hard liquor on the Sabbath.

This introductory architectural-geology tour begins not far from where Cap Streeter's weary vessel came to rest.

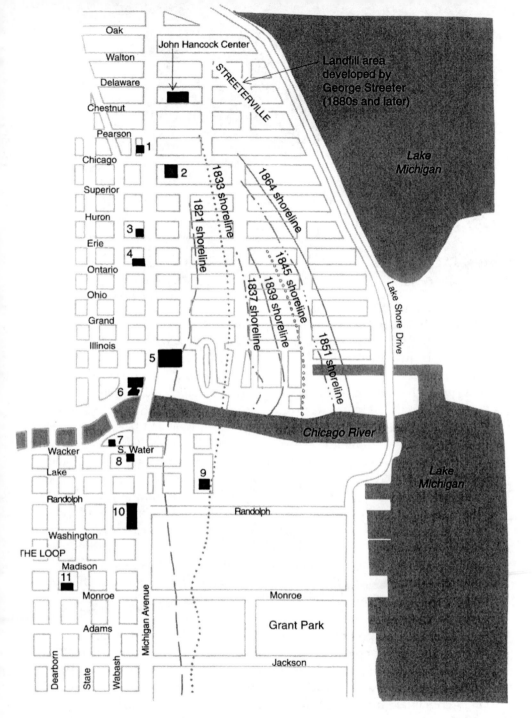

Downtown Chicago, with the eleven buildings visited on this tour shown in solid black and numbered. 1: The Chicago Water Tower; 2: The Olympia Center; 3: The Terra Museum of American Art; 4: The Woman's Athletic Club; 5: Chicago Tribune Tower; 6: The Wrigley Building; 7: The Seventeenth Church of Christ Scientist; 8: The Carbide and Carbon Building; 9: The Amoco Building; 10: The Chicago Cultural Center; 11: Inland Steel Building. —Modified from Bretz, 1955

A close-up of an old Chicagoland building faced with the Joliet-Lemont Sugar Run dolomite. The Sugar Run cladding is a particularly handsome material that often weathers to a buttery yellow. Unfortunately, it sometimes peels away in layers.

The Chicago Water Tower
(North Michigan Avenue at Chicago Avenue)

This modest but world-famous structure is Chicago's sacred totem pole. Because it survived the devastating 1871 fire, it became the city's symbol of determination and survivability. Designed by W. W. Boyington in a style that might be termed Amateur Gothic, it was completed in 1869. To the geologist familiar with this region's rock types, the handsome stone of the Water Tower and its associated pumping station is an old friend. Note its glowing yellowish tone: this is the trademark of the Silurian dolomite of the Sugar Run formation (often called, not quite accurately, limestone) quarried in the Lemont-Joliet area. Also offered commercially as Joliet or Athens marble—another misnomer—this stone was commonly used until Indiana's Bedford limestone supplanted it in the 1890s. The Bedford lacks the warm, golden tint of the local rock, but it does not exfoliate, or peel away in layers, the way the dolomite does. No one wants a building that sheds like a sheepdog. The Sugar Run dolomite was deposited some 420 million years ago, when Illinois was positioned below the equator and was covered with a warm, shallow sea.

The Olympia Center
(161 E. Chicago Avenue, half a block east of Michigan Avenue)

This 1984 example of postmodernism shows the latest major trend in architecture: after years of buildings fronted with steel and glass, ornamental stone is back with a vengeance. This structure is sheathed from head to toe in pink granite quarried in Sweden. The transatlantic seaborne-transportation costs for a stone shipment this massive must have been staggering: which perhaps is the We've-Got-It-and-We-Flaunt-It message inherent in this

Simple but effective contrast: the Olympia Center's pink Swedish granite near ground level. The upper panel has been polished to a high gloss; the lower panel remains unpolished and rough.

otherwise understated design. Some of this beautiful crystalline rock has been polished; and for simple but effective contrast, some of it has not. Take a hand lens or magnifying glass to this stone and examine the individual minerals. These small individual components slowly crystallized from a body of molten rock while it was still seated far underground.

The Terra Museum of American Art
(666 North Michigan Avenue)

The facade of this relatively small, new building is the best place in the city to see the glory of polished marble in the full sunlight. Marble is a metamorphic carbonate rock that has been a favorite of architects and sculptors for thousands of years. To be metamorphosed into high-quality, white or pale gray marble, the parent limestone must be almost pure calcite, and it must be subjected to high enough pressure and temperature to at least partially recrystallize the calcite. In many rocks, the constituent minerals are changed into something else during metamorphism; but calcite remains calcite. Note how the blue veins running through the white matrix add to its beauty. Veins in marble are usually composed of another carbonate mineral that is present in small quantities.

The Woman's Athletic Club
(626 North Michigan Avenue)

This 1920s-era building features a facing stone of particular beauty and elegance at the corner of North Michigan and Ontario. This veined, greenish black rock recalls the look of deep salt water on a cloudy day. It is a type of cladding (ornamental covering material) that architects call verde antique marble; to geologists, it's not a marble, but serpentine. If one exciting new theory is correct, you are looking at the rarest rock type of all: a highly

Work in progress on the Wrigley Building. The same natural forces that weather and weaken stone surfaces have attacked this structure's gleaming white cladding. Molded plastic replicas now replace the original terra-cotta tiles.

metamorphosed piece of the earth's mantle, the deep-seated zone that underlies our planet's relatively thin crust. The theory proposes that serpentine is formed in the chaotic setting of an oceanic spreading center, when a portion of the upper mantle comes in contact with seawater. Many millions of years later, a large assemblage of rocks, known as an ophiolite sequence, may be scraped up onto the leading edge of a continent when it collides with a volcanic-island arc. Serpentine is one small part of that ophiolite sequence. On both sides of North America—for example, in Vermont's Green Mountains and in California's Sierra Nevada—serpentine deposits are part of jumbled complexes that apparently began in the middle of ocean basins and ended up high in the continental highlands. No mean feat. The somber beauty of this serpentine reveals something more than our passion for using ornamental stone. It also demonstrates what remarkable changes the earth produces in the span of geologic time.

Chicago Tribune Tower
(435 North Michigan Avenue, at the Chicago River)

As every tourist wandering in this locale soon notices, the ground-level exterior of this Chicago landmark contains inset pieces of stone and building material taken from some of the world's most famous historical sites. Among the more geologically interesting specimens are a white granite or granitelike glacial erratic plucked from Boston's Bunker Hill, limestone from the Great Pyramid of Giza, and a chunk of coquina from Fort Maria, in St. Augustine, Florida. Coquina is a particularly eye-catching form of limestone, predominantly made of shell fragments and other hard remains of marine creatures, all cemented together into a crumbly rock with a texture that bears more than a passing resemblance to a granola bar. Of the three specimens cited here, only the Bunker Hill erratic is igneous in origin, rather than sedimentary. Look closely at its interlocking crystals, which long ago precipitated out of a molten slurry of magma far beneath the earth's surface.

The Wrigley Building
(410 North Michigan Avenue, at the Chicago River)

As the Water Tower is symbolic of the city's pertinacity, this gleaming mass of light and grace symbolizes Chicago's willingness to expend limitless energy to achieve its goals. At night, the structure is illuminated by banks of metal-halide lights that seem to outdo the sun. Originally, the building's cladding was glazed terra-cotta—baked clay covered with a glossy coating (in this case, a shimmering white). Still, the building's geological significance lies not in the terra-cotta itself but in how natural forces have worked upon it. For decades, the building's management had to deal with the fact that the tiles were developing hairline cracks in response to the great variations

The Wrigley Building's base, along the Chicago River. The lowest level (with the arched windows) is faced with Mississippian Bedford limestone; above it is the ornate glazed tiling.

in temperature. Rainwater seeping through the cracks and the mortar caused the supporting iron shelves to rust. The rust expanded, causing bigger cracks. In winter, water turned to ice and triggered another expansion process, frost wedging, the same process geologists often see dislodging rock in outcrops. By 1984, this assault of the elements prompted the Wrigley Company to begin replacing the terra-cotta piece by piece with plastic replicas.

The Seventeenth Church of Christ Scientist
(55 East Wacker Drive)

A close inspection of the exterior of this distinctive house of worship reveals that the facing stone is travertine, a striking rock type that contains banded patterns of spongy pockmarks. (The best place to see the banding is on the church's south side.) Travertine forms at the mouths of hot springs and in other zones of seepage, when calcium carbonate precipitates out of mineral-rich water. The most famous source of travertine is in the volcanic terrain around Tivoli, Italy; but it also occurs in small quantities in Illinois—for instance, in the bluff of the Mississippi River valley near Wolf Lake, where lime-saturated water has seeped through thick Pleistocene loess deposits.

Travertine, an unusual carbonate rock, has been a favorite of architects for centuries. Here on the Seventeenth Church of Christ Scientist, the pockmarks characteristic of this stone form a linear texture that runs perpendicular to the ground.

The Carbide and Carbon Building
(230 North Michigan Avenue)

What a darkly beautiful structure this gilt-trimmed Art Deco skyscraper is. At this stop, you can easily find an unusual limestone that is most often called black marble—an understandable gaff considering that marble is the metamorphic derivative of limestone. When you examine the main entrance, you'll notice that the black marble is offset by the mellowest of ornamental metals, bronze. One of the major sources of black marble is the Ordovician Chazy group—specifically, the Crown Point limestone, mined by the Vermont Marble Company in Isle la Motte, one of the islands in Lake Champlain. This stone, also nicknamed "Champlain Black" for obvious reasons, often contains the fossil snail *Maclurites magnus*. The existence of undeformed fossils is one indication that this handsome stone was not significantly altered by the powerful mountain-building forces that affected the Northeast in the Paleozoic era.

The Amoco Building
(200 East Randolph Street)

At eighty-two stories, this great white monolith is the second tallest building in Chicago. All too often architecture commentators have described it in scalding terms that could have melted the entire Wisconsinan ice sheet. One criticism is that it is too inhuman, too imperious, too impersonal. But to someone who in the early 1970s watched it rise on its exposed, provocative site—and who has since seen it float in the lake mist, seemingly detached from the earth—it appears that its lack of compromise is precisely its strongest point. If the architects have spurned it, I suggest that geologists, who have a greater tolerance of the inhuman, adopt it. Given the building's troubled history, it could use a sympathetic support group.

When first erected, the Amoco Building was sheathed in the renowned Carrara marble (real marble this time, and the very best). So much cladding was needed for a surface this huge that the project badly depleted the centuries-old, northern Italian quarries that once supplied Michelangelo with the raw material for his sculptures. But when the thin-cut marble slabs were shipped to the New World and installed on the Amoco Building's sides, things went horribly wrong. The marble, exposed to the worst of Chicago's climatic extremes, warped like waterlogged plywood. Imagine the collective look on the faces of the Amoco brass when they learned that the entire building would have to be resurfaced with thicker, more resistant white granite from North Carolina. It would not have been a good day to ask for a raise. The colossal resurfacing project took place in the early 1990s. I happened to walk by one breezy spring day when the newly arrived granite

The base of the soaring Amoco Building. North Carolina granite has replaced the shimmering Carrara marble that proved inadequate in this blustery, exposed site.

lay stacked on the street. The construction crew looked on in bafflement as I eagerly examined the stone with my hand lens. You can easily do the same at any point within reach on the building's exterior. Keep in mind that this rock, which imparts a much duller luster than the gleaming Carrara marble did, was emplaced in the heart of the ancient Appalachians late in the Paleozoic. In that era, eastern North America was the continent's leading edge, and over the course of many millions of years it collided with several other landmasses, including northwestern Africa.

The Chicago Cultural Center
(78 East Washington Street)

For many years this noble edifice was the main branch of the Chicago Public Library. Its exterior combines pale granite with the Midwest's most famous and most widely used sedimentary cladding, the Bedford limestone. Unlike the Silurian Sugar Run dolomite used at an earlier phase in Chicago's development, this carbonate rock often has a cold, cement-gray tint. What makes it wildly popular is that it is easy to cut and carve, and very durable. Reputedly, it even gains strength as it ages; and it does not stain when subjected to weathering. An ironic result of this stone's evident superiority is that venerable buildings sheathed in it often look like two-year-old imitations of venerable buildings. The Bedford is quarried in southern Indiana;

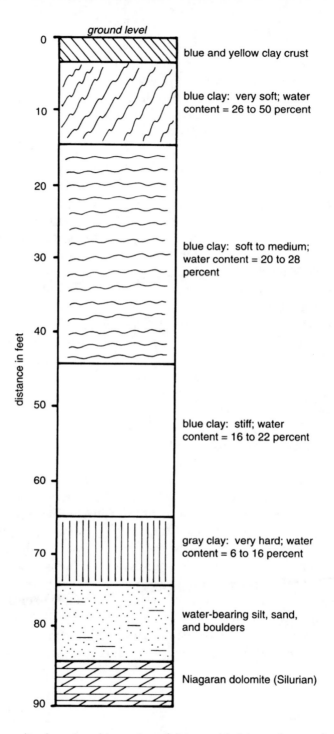

ground level

0 — blue and yellow clay crust

blue clay: very soft; water
content = 26 to 50 percent
10 —

20 —

blue clay: soft to medium;
30 — water content = 20 to 28
percent

40 —

50 —

blue clay: stiff; water
content = 16 to 22 percent

60 —

70 — gray clay: very hard; water
content = 6 to 16 percent

water-bearing silt, sand,
80 — and boulders

Niagaran dolomite (Silurian)

90 —

distance in feet

Generalized stratigraphic section of the unstable lake sediments that
underlie the Inland Steel Building. To anchor the building, steel pilings
were driven through nearly 85 feet of silt and clay to the more solid
Silurian Niagaran dolomite. —Illinois State Geological Survey

it is part of the same Mississippian Salem formation that outcrops in south-western Illinois. Geologists call this type of limestone biocalcarenite. It is made up of tiny fossil fragments cemented in a matrix of calcite.

If you walk into the Cultural Center's Washington Street entrance, you'll also find a superb example of the Carrara marble now so vividly absent from the Amoco Building. The splendid main staircase in front of you is made of this premium ornamental stone. When newly cut, it has a sugary texture; if you run your fingertips over its highly polished surface here, you may be able to detect it. You may also spot some of the interlocking calcite crystals that reveal that this stone was once unmetamorphosed limestone. A distinctly different marble, quarried in Ireland, borders the mosaics of the staircase's railing. It's just the color stone from the Emerald Isle should be: a deep, lustrous green.

Inland Steel Building
(30 West Monroe Street)

No baby boomer can look on this stainless-steel gem of an office building without feeling a pang for the brave new world of 1950s modernism. This green-and-silver masterpiece, adored by practically everyone with an interest in architecture, may seem distinctly ungeological. But here, obviously, the geology lesson lies not in the building's exterior but in how its engineers coped with the treacherous lake sediments described at the head of the essay. The Inland Steel Building was the first structure to use what has since proven to be the most effective anchoring technique: steel pilings driven all the way through the thick blanket of silt and clay to the much firmer Silurian dolomite bedrock. Here, the pilings had to extend 85 feet straight down before they met the solid pre-Pleistocene surface. In earlier decades, builders tried some less successful stabilization schemes. One was simply to spread out the weight of the building as broadly as possible, on a raftlike base. Caissons—first used in the last decade of the nineteenth century, for the late, lamented Chicago Stock Exchange—were another partial solution. They consisted of deep holes lined with timber and filled with concrete. They attached the buildings' foundations not to distant bedrock but to a stiffer layer of clay known as hardpan. Despite all the care, settling remained such a common problem that many building entrances were placed higher than normal, to compensate in advance for the sinking that was likely to follow.

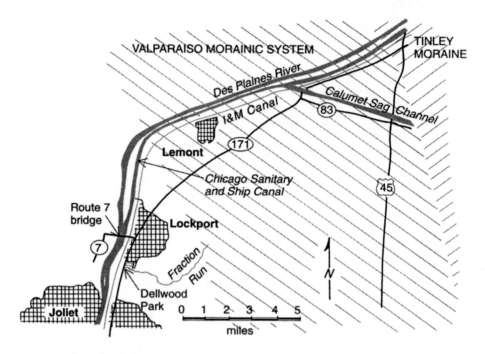

Lockport and vicinity. The geologic tour begins on the Illinois 7 bridge over the Des Plaines River north of Joliet.

— 13 —
Exploring the Lower Des Plaines
LOCKPORT AND LEMONT
Will County

For the moving of large masses of rock, the most powerful engines without doubt which nature employs are the glaciers. . . . These fragments they gradually transport to their utmost boundaries, where a formidable wall ascertains their magnitude, and attests the force, of the great engine by which it was erected. The immense quantity and size of the rocks thus transported, have been remarked with astonishment by every observer.

—John Playfair, *Illustrations of the Huttonian Theory of the Earth,* 1802

The land southwest of downtown Chicago is proof enough that Playfair's dramatic assertion is no exaggeration. There stand two formidable walls, the Valparaiso morainic system and the Tinley moraine, demonstrating how effective a transportation system the Wisconsinan glacier was. Now the Des Plaines River and other waterways flow between and even through these high walls of glacial sediments. They, too, have been impressive transportation systems, but of a different order. Instead of hefting immense quantities of rock debris, they have helped human beings distribute the goods of countless forms of commerce.

The best way to start your exploration of this area is to enter Lockport from the west, across the Route 7 bridge. The view from this span is remarkable: there is not a single main stream in its valley, as one would expect, but three parallel waterways with no high ground between them. The only natural river here is the one the American Indians and the French voyageurs knew and used so well: the Des Plaines. It retains some semblance of its old self; and in the spring thaw it still swells over its banks and rushes briskly along—a faint echo of the great Lake Chicago and Lake Nipissing floods that used this same valley periodically from about 14,500 to 4,000 years ago. The suspicious straightness of the other two waterways tells even the untrained eye that they are man-made features. The larger one, often busy with tugs and their tows, is the Chicago Sanitary and Ship Canal. The smaller one, now devoid of the mule-guided wooden barges that once drifted silently

The massive Niagaran dolomite in a cut facing the Illinois and Michigan Canal, in Dellwood Park. The dark blotches may be air-pollution staining of chert nodules in the dolomite.

by, is the Illinois and Michigan Canal. Take a moment to speculate: why were these channels built so close to a preexisting river? The main reason is that it was much easier to dig a channel in a preexisting valley than to excavate through the surrounding high, morainal ground. The preexisting river also provides a convenient source of water for the canal.

Lockport has more character per square inch than any other river settlement in Illinois. The prevalence of old buildings faced with warm yellow Silurian dolomite from the Sugar Run formation makes it tempting to think that the town has risen conformably out of the rock that underlies it. On Lockport's south side, along Route 171, you'll find lovely little Dellwood Park, a site that appeals to the geologist and historian alike. Proceed to its southern end and walk down the steps to the dam and tunnel where Fraction Run, the local creek, pours into the languid waters of the Illinois and Michigan Canal. The I&M may not be an imposing sight, but it would be difficult to overestimate the role this slender strip of water played in the development of the Chicago region and, in fact, of the nation.

To the mathematically minded person, the canal is a game of sixes: it was built 60 feet wide at the surface, 36 feet wide at its bottom, and 6 feet deep. The construction of the channel, which for the most part involved excavating through unconsolidated Pleistocene sediments rather than through large portions of the deeper-seated bedrock, proved to be an unexpectedly perilous and financially shaky undertaking that was fraught with much delay.

The lives of hundreds of workers—most of them newly arrived Irish immigrants—were lost before the I&M finally opened for business in 1848, twelve years after the work had begun. Once in operation, this 96-mile channel linked the struggling, sparsely populated settlement of Chicago with the Illinois River and the markets of the lower Mississippi. Partly as a result of this engineering feat, the Windy City beat out St. Louis in the competition to be North America's greatest inland commercial and transportation hub. Ironically, the I&M was quickly eclipsed by the railroads, which also used the low-cost route along the Des Plaines River corridor. Then in 1900 the completion of the much roomier and deeper Chicago Sanitary and Ship Canal (dug deep into bedrock with a much more powerful stream-driven technology) put an end to the I&M's usefulness. This most important Illinois landmark soon fell into a state of trash-filled disrepair, but in 1984 it took on a new role as an educational and recreational resource. Congress declared it the country's first National Heritage Corridor. You can learn more about the I&M's historic sites, bike trails, and special events by visiting the canal's old headquarters building in downtown Lockport.

After you've examined the canal, take a close look at the cliff of Silurian-period Niagaran series dolomite that fronts the trail. You may notice dark, tarlike marks on the rock. In other locales, where these strata contain ancient reefs, accumulations of bitumen, otherwise known as asphalt, are common. This bitumen is a chemical residue of organisms that once inhabited the reefs. Here, however, the lack of a reef structure suggests that the staining is the result of some other factor. Chert nodules in exposed Niagaran outcrops elsewhere in Chicagoland have been blackened by air pollution. Perhaps some similarly unnatural process accounts for the spotting here.

Walk back through the tunnel and into the small gorge that holds Fraction Run and the park's lower level. The excellent layered cliff exposures give you an even better feel for the Niagaran series. This is part of the same stratigraphic section that is revealed extensively at the great Thornton Quarry (see essay 15), and that also outcrops in Wisconsin's Door Peninsula, and at Niagara Falls, far to the east.

When Lake Chicago breached its morainal dams and spewed forth its floodwaters near the end of the Pleistocene, the Des Plaines River was cut down much deeper than before; but tributaries such as this one found themselves still perched on the old, higher surface. Eventually, they adjusted to the new base level by cutting downward, too. But in the case of Fraction Run, the cutting had to be done through highly resistant Niagaran dolomite. As you see here, the creek managed to chew its way through, but in the process it was confined to a narrow valley.

Silurian Niagaran series dolomite of Dellwood Park's southern cliff face.

Route 171 north of Lockport passes near Lemont, once an important source of building stone, and up onto the lofty ground of the Valparaiso morainic system. Anyone caught complaining about Illinois's flatness should be sentenced to ride up this incline on a bicycle with only one gear. This massive barrier of glacial till, together with the separate Tinley moraine a little to the east, was formed slightly less than 15,000 years ago, during several different episodes when the Wisconsinan ice sheet paused long enough to dump huge masses of rock debris along continuous fronts. To see the two places where Lake Chicago managed to cut through the Valparaiso and Tinley ridges, continue northeast on Route 171 to Willow Springs—the vicinity of the northern outlet—and then head south to the junction of Routes 45 and 83—the southern outlet's locale. These ancient spillways of Lake Chicago now hold the Sanitary and Ship Canal and the Calumet Sag Channel. Here, human engineering has once again mimicked natural forces and features, taking advantage of the low ground and gaps in the moraines. In meek imitation of the long-vanished days of the Lake Chicago floods, today's rain runoff from the metropolitan region, and even some of Lake Michigan's waters, have been diverted southwestward, where they are destined to travel to the Gulf of Mexico.

110

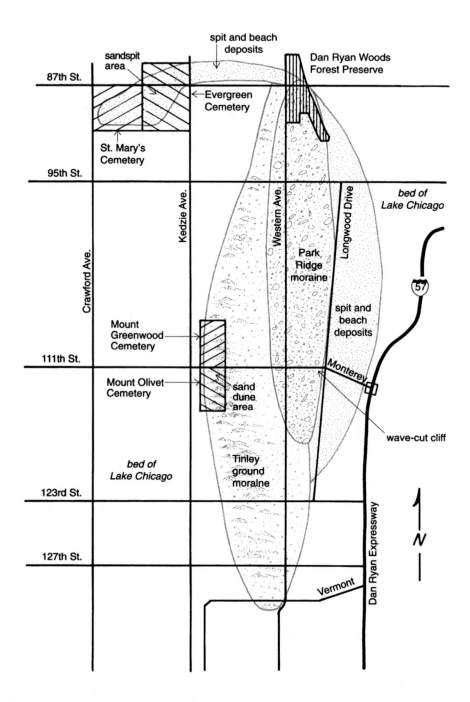

Blue Island, a prominent ridge of morainal deposits on Chicago's south side.

— 14 —
Blue Island Bracketed
CHICAGO'S SOUTH SIDE
Cook County

Glacial epochs are great things, but they are vague—vague.
—Mark Twain

The recent past can be foggier than the distant past. It is one of the quirks of geology that in some ways we know more about the Illinois of the Coal Age (approximately 300 million years ago) than the Illinois of the Ice Age (just a few millennia ago). When you try to piece together the chronology of the Pleistocene events that affected the Chicago region, you quickly find that much of the record has been obliterated: later glaciers often destroyed evidence of earlier ones, and whatever evidence survives sometimes points to conflicting causes. This can lead to a tangled mass of theories and interpretations that may never be fully resolved. We may never know exactly when the swollen surface of ancestral Lake Chicago reached its high-water marks, or precisely where the Wisconsinan ice sheet was located in any particular century. A certain measure of uncertainty is the price we pay for being interested in the earth's imperfect record.

Some ancient landmarks in northeastern Illinois, though, are straightforward in their origins. Their interpretation varies little if at all from one geologist to the next. One such agreeable feature is well known to every resident of Chicago's south side: the prominent ridge called Blue Island. It runs from the vicinity of Dan Ryan Woods, on the north, to the suburb that bears its name, on the south.

For once, a popular name is accurate. In the closing years of the Ice Age, Blue Island was an island—a narrow rampart in the high waters of Lake Chicago—when most of the rest of the metropolitan area was deep under the waves. Today, Blue Islanders might be surprised to learn that their home turf has more in common with the rolling terrain of the North Shore than it has to do with the city proper. While its southern and western flanks are composed of Tinley ground moraine, the most impressive part of the ridge is a detached portion of the Park Ridge moraine, the westernmost member of the Lake Border morainic system. It is composed of glacial till dumped by an ice sheet that was still moving forward from its rear, but was also

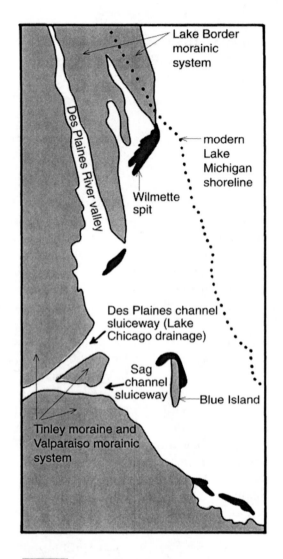

Lake Border
morainic
system

modern
Lake
Michigan
shoreline

Wilmette
spit

Des Plaines channel
sluiceway (Lake
Chicago drainage)

Des Plaines River valley

Sag
channel
sluiceway

Blue Island

Tinley moraine and
Valparaiso morainic
system

High morainal ground

Sandspits built in nearshore water (submerged at Glenwood high-water phases)

*The Windy City area in the days of ancient Lake Chicago. At its highest
stages, the lake stood about 60 feet higher than it does now. The Des
Plaines and Sag sluiceways were outlets that allowed the meltwater
from the still-extant ice sheet to the north to drain into the Illinois/
Mississippi River system.* —Modified from Bretz, 1955

Buildings and pavement notwithstanding, this section of Blue Island's wave-cut cliff, at 111th Street and Longwood Drive, is still a dramatic dropoff. At the highest Glenwood phases of Lake Chicago, the foot of this rise would have been under water.

melting enough at the front for the leading edge to remain in one location for an extended period. Along its crest, Blue Island is approximately 660 feet above mean sea level. The highest mark hit by Lake Chicago, during the Glenwood stages, was 640 feet.

In modern times, the Blue Island moraine is a bustling, built-up area. Nevertheless, it's an easy task to track down some of the most geologically significant details—even if they're paved over. One of my favorite viewpoints is located at 111th Street and Longwood Drive, a few blocks west of the ramp off the Dan Ryan Expressway (Interstate 57). At this intersection, you are on the Glenwood level of Lake Chicago; abruptly to your west rises a steep incline, a cliff that was cut by the Lake Chicago surf at it highest extension. If the din of car horns and passing buses permits, picture yourself alone on this windy strand 14,000 years ago. You hear the cry of seagulls and the hissing of waves; but the sounds of human civilization have not yet been heard on earth. A ten-minute walk uphill gives you a commanding view unhindered by buildings or fence lines. The limitless blue water, which until recently was locked up in the Michigan, Saginaw, and Erie lobes of the Wisconsinan ice sheet, surrounds you. In fourteen millennia, instead of being one of the most isolated spots imaginable, this will be a busy place.

115

If you continue up the hill, due west on 111th Street, you will soon come to Mount Greenwood and Mount Olivet Cemeteries. Despite the effects of the landscaping, they still show the subtle highs and lows of sand-dune terrain. The prevailing winds mounded and molded these forms here, on the old island's western side. Illinois stratigraphers classify these dune deposits as the Parkland sand—a generic term that also refers to other windblown sand of Pleistocene and Holocene origin. In case you're wondering, the presence of cemeteries here is no coincidence. Gravediggers much prefer to do their work in this kind of dry, yielding, well-drained soil.

At the risk of sounding obsessive, I recommend you make a short stop at one more cemetery. Head up the moraine's western side, along Western Avenue. At the northern end of the ridge, turn left onto 87th Street. You probably won't notice it, but you are heading out onto an ancient sandspit that curls like a giant talon to the west of the moraine. This spit was formed by Lake Chicago's longshore currents, which dropped their load of sediments around Blue Island's tip.

The Pleistocene dune complex on Blue Island's western side is now largely used for burial grounds. This view of Mount Greenwood Cemetery shows the eastern slope and top of one of the dunes.

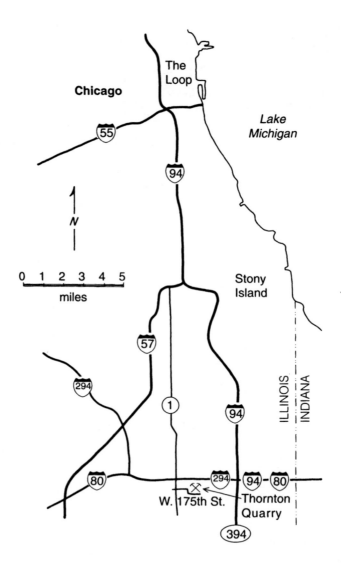

Location of Thornton Quarry, south of Chicago.

— 15 —
A Deep Look at the Distant Past
THORNTON QUARRY
Cook County

Nor let it be thought that any devotion to paleontology—to the "stocks and stones" of the sneerers at science—will ever lessen our love for the fresh and beautiful in existing nature. . . . As the existing throws new light on the extinct, so the extinct adds fresh interest to the existing; and thus, to the paleontologist, the study of life becomes not only a more exciting pursuit, but a higher and more ennobling theme.

—David Page, *The Past and
Present Life of the Globe*, 1861

As this Victorian-era science writer suggests, the paleontologist has a certain perspective the rest of us should envy. The person who studies the record of ancient life as revealed in the fossil record has tapped into something larger than a collection of arcane trivia about long-vanished plants and animals with strange Latin names. The paleontologist has gained the power to breathe life into the world of many millions of years ago—to explain, however incompletely, a successful system of life that flourished long before ours came to be. The reconstruction of the vast span of life's history seethes with poetic content.

For several decades, one of the most engaging disciplines of earth science has been paleoecology, the study not just of separate fossil life-forms but of their interrelationships, and of the environments they inhabited. It was in the Chicago region that paleoecology was largely born. In the first half of the twentieth century, specialists affiliated with the Illinois State Geological Survey, the University of Chicago, and several other institutions began a detailed study of the superb organic-reef complexes they found preserved in Middle Silurian carbonate rocks of the Niagaran series. When those reefs were built by corals and many other kinds of sea creatures some 425 million years ago, Chicagoland was covered in a shallow, clear, tropical or subtropical sea. Further, reconstructions of the movement of continents over geologic time suggest that Silurian North America was situated south of the equator, thousands of miles south of its present position. In those days, the dry land nearest northern Illinois stood far to the east, where the

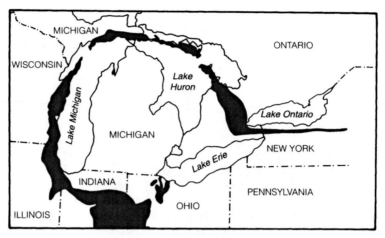

The Silurian carbonate strata of the Niagaran series (solid black) ring the Great Lakes region. The strata are visible in many places, including Wisconsin's Door County Peninsula, Niagara Falls, and Chicagoland's huge Thornton Quarry. —Modified from Bretz, 1955

once-lofty Taconian Mountains, which had risen in Ordovician time, were rapidly being laid low by the inexorable forces of erosion.

Illinois's reefs are only one part of a much larger complex that fringes the great sedimentary basin centered on Michigan's Lower Peninsula. This complex has been compared to an earlier version of Australia's Great Barrier Reef. Taken individually, these structures have certain telltale features that help geologists map their structure and extent. They range in size from the dimensions of a two-car garage to roughly the diameter of Chicago's Loop, but they have a consistent domelike cross-sectional profile. In the center is the dense core rock—usually pure, unlayered, and fossiliferous dolomite. Not surprisingly, the commonest types of organisms preserved in the cores are sedentary reef-building kinds: corals, calcareous algae, and stromatoporids—spongelike, lime-secreting animals. On the core's flanks lie strata that tilt downward and away from the center at anything from 50 to 5 degrees. These dipping strata are a rare example of sedimentary rock layers that were not originally laid down in a horizontal position, because their sediments came to rest on the reef's sloping sides. But eventually, outward from the reef core, the layers have become essentially horizontal. This portion is called the inter-reef zone. The fossil record here speaks of a sparser and less robust population that includes sponges. To the geologist and quarry

120

A chunk of massive Niagaran dolomite, a geological trademark of the Thornton Quarry and the Chicago region in general. The black bitumen that dapples the rock is the residue of marine creatures that thrived in Illinois's inland seas some 420 million years ago.

operator alike, these sponges played a particularly significant role. The silica hard parts they secreted were later transformed into the chert nodules that are fairly common in Silurian carbonate rocks—or so a current, well-accepted theory holds. Chert, an extremely hard mineral, may have been a favorite tool-forming material for American Indians, but it is unsuitable for the production of concrete. Also, it can be the bane of mechanical rock crushers. If it is present in any quantity, it can produce an effect roughly equivalent to dropping steel ball bearings into a kitchen blender.

In the Chicago area, an impressive Silurian reef is now the site of one of the world's largest quarries devoted to the production of crushed stone. This famous facility, one of Chicago's most breathtaking landmarks, is the Thornton Quarry: the huge rock-walled pit through which Interstate 80/294 and its puny human traffic pass. Geologists have estimated that at its maximum extension the main Thornton reef may have been 600 feet high—the height of a fifty-story office building—and well over a mile wide in diameter at its base. This fact makes us realize that Chicago is by no means

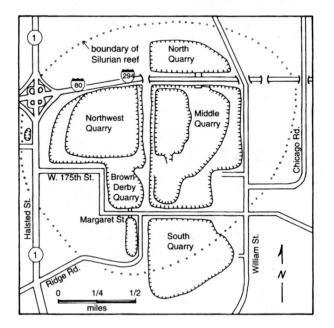

*The Thornton
Quarry complex.*
—Modified from Mikulic
and Kluessendorf, 1994

*Three Silurian fossils found in the Thornton reef, from left to right: the
trilobite* Calymene *and the corals* Halysites *and* Favosites. —Illinois State
Geological Survey

the first great collection of construction-crazy creatures to rise in these parts. When after millennia of slow deposition the Silurian sea finally withdrew, this reef was exposed to many millions of years of erosion. Finally, its uppermost 300 feet was planed away, but because its pure dolomite is more resistant to disintegration than the other carbonate rocks nearby, the reef formed a mound or low hill. (In geo-jargon, this kind of prominence formed by an exhumed reef is called a klint; the plural of this originally Scandinavian term takes an -ar ending. Of Chicagoland's several klintar, Thornton and Stony Island are the best known.)

Before the advent of the Ice Age, the summit of the Thornton klint stood perhaps 50 feet above its surroundings. During the Pleistocene, however, it was draped in a mantle of unconsolidated sediments that reduced its relief to about half that height. The first quarrying here began as far back as 1837, though it was a modest operation by current standards. Today, the quarry covers more than 500 acres; its six separate, steep-walled sections, in some places more than 200 feet deep, are linked by huge arched tunnelways. Paleontologists who have collected here have found the remains of a diverse array of marine organisms as the rock removal has gone on, decade after decade. Veteran fossil collectors will recognize most of this abbreviated cast list of Thornton characters: the corals *Halysites* and *Favosites;* the trilobites *Calymene* and *Bumastus;* and the brachiopods *Eospirifer, Kikidium,* and *Wilsonella.*

The quarry operators plan to keep the Thornton facility active until they reach the 400-foot depth. By the mid-1990s, roughly three-quarters of the main reef structure had been excavated. In a century and a half— the blink of an eye, in geologic terms—human beings have outstripped whole geologic eras' worth of natural erosion and have created this monument to our drive to build, organize, and survive. Should the klint have been saved and more gingerly excavated, the same way archaeological sites are now being set aside and studied? It is too late to ask the question. Still, we need to remember that it was scientists and educators who advocated the preservation of a portion of the Stony Island klint as a state park, even though the state legislature did not support the proposal.

Public access to the Thornton Quarry is limited to field trips staged by the Chicago Academy of Sciences once or twice a year. But even if you don't get inside the facility, you can get a sweeping view of the quarry and its workings by taking Route 1 south from Interstate 80 and then turning east onto West 175th Street and then onto Margaret Street. In this stretch you may wish to stop along the roadside and peer through the chain-link fence at the massive pits and their exposures. Be careful not to hinder passing

A section of the Thornton Quarry, from the south. One of the arched tunnels that connect the quarry sections stands at left center. A huge pile of light-toned crushed stone is at lower right. Newer and lower levels of excavation are visible in the right background.

vehicles, and do not trespass on quarry property. One of the best views is from Interstate 80; keep in mind, though, that stopping along the shoulder for anything less than an emergency is illegal. Few state troopers, however well seasoned in the odd ways of public behavior, would consider an unrequited urge to stare at an ancient reef a life-threatening scenario. What can be life-threatening, however, is a driver who zips through the quarry with both eyes glued to its inner recesses. Avoid the temptation. Get some-one else to drive, so you can view the passing spectacle from the passenger seat. When you make the transit this way, don't forget to look at the do-lomite strata exposed in the interstate's roadcuts in the vicinity of the quarry. As you pass from one side of the reef to the other, you'll see how the angle and direction of these beds change in accordance with the inner structure of this geologic marvel.

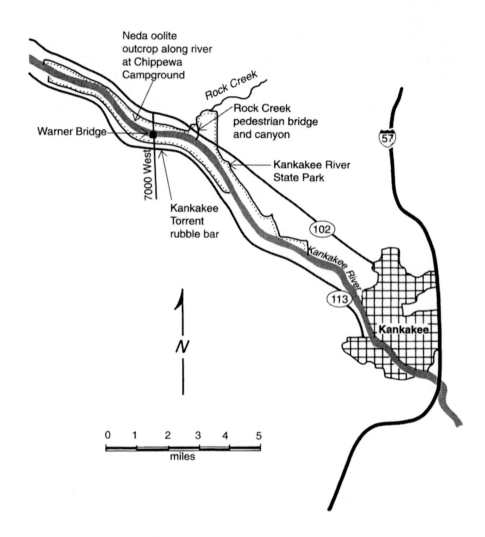

Kankakee River State Park and environs.

— 16 —
In the Path of the Kankakee Torrent
KANKAKEE RIVER STATE PARK AND ENVIRONS
Kankakee County

Destruction and disorder of the elements,
Which struck a world to chaos, as a chaos
Subsiding has struck out a world: such things,
Though rare in time, are frequent in eternity.
— George Gordon,
Lord Byron, *Cain: A Mystery*

When two hundred years ago the geological theorist James Hutton insisted on a world of mostly steady processes operating over great spans of time, he did so partially to demonstrate the earth's great antiquity. Ever since, his grand revision of the earth's age has been an unacceptable concept to people who insist that the Grand Canyon was formed in a few days by the draining of Noah's flood, and that the plant kingdom was created just a few thousand years ago, before the birth of the sun. Still, Hutton's idea— that ours is only the latest and slimmest chapter in the massive volume of geologic time—has been one of the greatest episodes of consciousness-raising in all of human history. It has been proven true again and again.

This is not to say, however, that every argument used by Hutton has survived the scrutiny of science. His downplaying of major catastrophic episodes in geologic history was unnecessary and anything but accurate—even at a time when catastrophe was usually considered synonymous with supernatural intervention. While much of our planet's record has been the result of slow, sustained effects, there have been short periods when rapid, natural change has been the driving force over large areas of the earth's surface. Residents of the Pacific Northwest who witnessed the regional impact of the 1980 eruption of Mount St. Helens would not dispute this; nor would the early-nineteenth-century settlers in Missouri and Tennessee who saw how the great New Madrid earthquakes altered the bed of the Mississippi River and the uplands around it. Further back in geologic time there are even more impressive examples of natural, widespread catastrophes. One of these produced its greatest effect in northeastern Illinois, and it was a spectacular happening by anyone's standards. Geologists have named this first-rate demonstration of sudden and dramatic change the Kankakee Torrent.

127

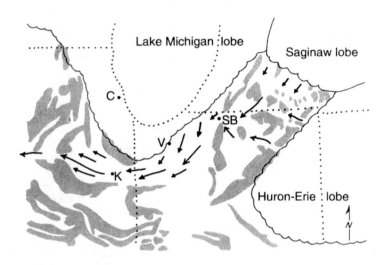

An interpretation of how the waters of the Kankakee Torrent made their way from the melting ice sheets to the Illinois River valley. The gray areas are moraines; the dotted lines show state boundaries. C=Chicago, K=Kankakee, V=Valparaiso, and SB=South Bend. —Illinois State Geological Survey

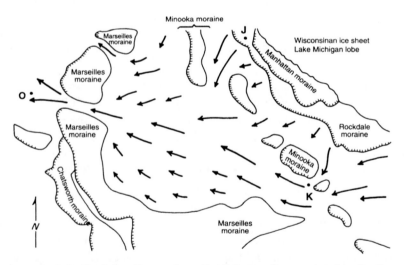

A close-up of the Kankakee Torrent from Kankakee to Ottawa. J=Joliet, O=Ottawa, and K=Kankakee. The water velocity, represented here by arrow length, was probably highest in the central portion of the channel—the path now occupied by the modern Kankakee River. —Illinois State Geological Survey

The Kankakee Torrent was an immense flood of glacial meltwater approximately 15,500 years ago. It flowed in the same area and direction that the much smaller and tamer Kankakee River does today. Other major floods in northeastern and central Illinois came later: huge volumes of water were released in distinct pulses at least several times late in the Wisconsinan stage of the Pleistocene. By present-day standards, these inundations would have been unimaginable. It is even possible that they not only affected the northern two-thirds of Illinois but also changed the courses of the Mississippi, Ohio, and Cumberland Rivers at the state's southern tip. (The reasons for this are discussed in essay 36.)

To see some of the lasting effects of the Kankakee Torrent, pay a visit to Kankakee River State Park and the surrounding countryside. Before entering the park proper, take a look at some of the most dramatic evidence of the flood's force. This is in the form of large rubble bars, situated on the southern side of the modern Kankakee River, along Route 113 in the vicinity of the Warner Bridge. These streamlined, elongate ridges parallel the river.

A Kankakee Torrent rubble bar seen in cross section along 7000 West. Despite the vegetation cover, you can spot chunks of white Silurian dolomite that were dumped here by the surging glacial meltwaters. The floodplain of the present-day Kankakee River lies just to the north, in the background.

Rock Creek Canyon, in Kankakee River State Park. The hard, cliff-forming strata are dolomites of the Silurian Joliet formation.

They might be mistaken for low bluffs or sand dunes, and there are plenty of the latter a little downstream. But if you turn south on 7000 West at its junction with Route 113, you'll get a glimpse of what these rubble bars are really made of. Close to the junction, at the first rise, a bar is exposed in cross section. The roadcuts are now largely grassed over, but you can still see rock fragments, most usually of Silurian Niagaran series dolomite, poking through the vegetation. Fast-moving rivers can carry sand and even gravel; but for water to pry large chunks of stone from the bedrock, transport them downstream, and fashion them into large bars, it must have extraordinary velocity and depth. This site, therefore, gives you a vivid picture of the torrent's carrying power.

The low-lying land in this area, including the broad bar-dotted plain on the river's southern side, was part of the Kankakee Torrent channel. In Kankakee County, the lowland remained a gigantic wetland long after the torrent subsided. Well into the nineteenth century, the locale was synonymous with impassable, mosquito-infested morasses. The rate of settlement here lagged far behind other portions of the state, despite the proximity of Chicago's skyrocketing population. Only when an extensive cooperative drainage system was imposed on the area did Illinois simultaneously gain much farmland and lose one of its most priceless, if most misunderstood, ecosystems.

The Kankakee Torrent also did much to create one of Illinois's most scenic attributes: hanging tributaries and their waterfalls. Normally, tributaries keep pace fairly well with the base level established by their parent rivers: from headwaters to mouth, their streambeds describe a gentle curve of increasingly shallow slope. But after the pulses of the torrent dwindled and the energy of the remaining water flow was focused on the channel of the Kankakee River, the main stream cut down with a speed its feeders could not match. To this day, the side streams are not well adjusted to the Kankakee's lower level. This is particularly apparent in the main section of the state park, where Rock Creek flows down into the main valley from the north. When you take the pedestrian suspension bridge across Rock Creek Canyon, you'll see that this tributary has here sliced its way down through the Joliet formation dolomites of the Silurian-period Niagaran series. (These are the same rock strata that form the base of the great Thornton reef described in essay 15.) At this spot, Rock Creek is very near its confluence with the Kankakee. If you take the trail up the canyon bluff, however, you will eventually come to a waterfall that marks the farthest point to which the creek has been able to lower itself in the past ten millennia or so. According to the Illinois State Geological Survey, the waterfall is working its way headward at an average of 3 inches a year. If your great-grandchildren visit this site near the dawning of the twenty-second century, they'll have to walk an extra 25 feet northward to reach the edge of the falls. The main story in Kankakee River country may be one of catastrophism; but Hutton's ghost can still take pride in all the sideshows, such as this one, that demonstrate the slow but inexorable change worked by water flowing over stone.

Kankakee River State Park has one other geologic point of interest quite separate from its Pleistocene and Holocene landforms. At the riverfront outcrop below Chippewa Campground (a little west of the Warner Bridge), there is a rare exposure of the Ordovician rock that underlies the Silurian strata. To reach it, park at the campground's western end and carefully descend the steep slope toward water's edge. Most of the rock exposed here is dolomite of the Kankakee formation, which dates from the Silurian. But at the bottom of the outcrop, you should find a particularly interesting reddish brown stone that once was called the Neda iron ore but which is now better known as the Neda oolite. This formation is the uppermost part of the Maquoketa group, and of the entire Ordovician system in Illinois. You'd be hard-pressed to find a good exposure of it elsewhere in the state.

A disconformity between Silurian and Ordovician rocks. The Ordovician Neda oolite is on the bottom, extending half again as high as the cassette recorder. The more massive carbonate rocks above it are part of the younger, Silurian Kankakee formation. Note the vertical joint, or fracture, in the lower part of the Kankakee formation, just right of center.

What makes the Neda remarkable is its composition. Using a hand lens, you'll see that it is an odd sort of shale with a striking pattern of tiny grains of iron minerals—either hematite or goethite. If the rock is weathered, the iron-oxide mineral limonite will give the rock's surface a yellowish brown cast. The Neda oolite formed in unusual conditions. The waters of its birth, the Upper Ordovician sea that covered this area, were saturated with iron for some reason we cannot yet divine. When the iron precipitated out into solid form, it coated tiny bits of other matter to make these sandlike grains. Stratigraphers who study this elusive formation have found crossbedding patterns in it, which indicate the rock was deposited in a shallow-water, high-energy environment, well within the work of the waves.

When you locate the Neda oolite, trace it upward a few inches to its contact with the much more massive, nongrainy Kankakee dolomite. What do you notice about the boundary between the two formations? Is the boundary smoothly transitional, or is there a distinct break, even though the two units fit snugly together? The second option is the true one: you're looking at a disconformity, a significant gap in time between two parallel sets of strata. Between the points when the deposition of the Neda oolite ended and the deposition of the Kankakee began, the continental sea that covered this area withdrew when the worldwide sea level dropped as an ice sheet developed in the southern polar region. The upper part of the Neda was exposed to erosion for an extended period. Fortunately for us, the early Silurian sea moved in to bury the bottom portion of it in new sediments. As a result, this delicately textured rock, so appealing to the inquisitive human eye, was not completely erased from the record of the earth's crust.

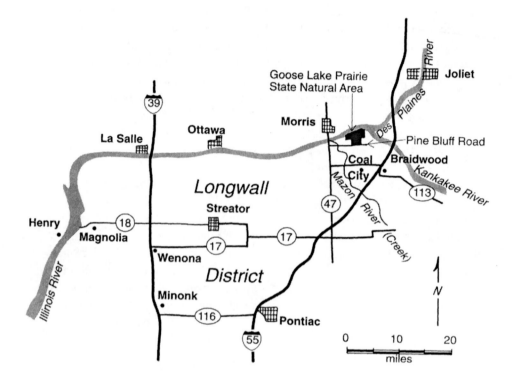

Mazon Creek and the Longwall District.

‒ 17 ‒
The Realm of Fossil Forests
THE MAZON CREEK AREA AND THE LONGWALL DISTRICT
Will County to Woodford County

In this time when the lands were raised above the waters, they were covered with great trees and vegetation of all kinds. The world-wide sea was . . . the universal receptacle for all that which broke away from the lands which rose above it. The quantity of vegetable products and remains . . . is too great to be described.

—George-Louis Leclerc, Comte de Buffon,
Époques de la nature, 1807

To someone more interested in mining or economics than in fossils, this essay might better be entitled "The Realm of the Colchester Coal," because without this seam of black, compressed, hardened peat the history of the Prairie State would have been much different. But the fossil forests must have their due; they have fired the imagination of thousands upon thousands of Illinois schoolchildren and adults. In this state, it is almost impossible not to know someone who has his or her own collection of plant-leaf impressions gathered in the strip-mined land south of the upper Illinois River.

If you drive on Interstate 55 between Braidwood and Gardner, you will cross what the official Illinois road map calls the Mazon River; but if you sneeze at the wrong instant, you'll miss it. Geologists and paleobotanists know it by the more appropriate name of Mazon Creek. Despite its diminutive size, it holds a place of honor in the hearts of students of ancient plant life. It is in the mined-out land surrounding this humble stream that some of the most extensive remains of Pennsylvanian-period vegetation have been unearthed. Known as the Mazon Creek Biota, this assemblage of fossilized plants and animals has given one generation of scientists after another detailed information about the verdant living world of 300 million years ago.

From a paleobotanist's perspective, the Pennsylvanian rocks exposed in this region and elsewhere hold two types of treasures. The first, the one highly prized by amateur collectors, is the ironstone concretions that often contain impressions of plant leaves, stems, and reproductive structures, as well as the remains of animals. One of those animals, the bizarre soft-bodied Tully Monster, has managed to edge out all the spectacular coal-swamp plants to be declared the Illinois state fossil. Illinois State Geological Survey scientist Donald Mikulic—whom I might respectfully describe in this matter as a fossil-animal-rights advocate—has pointed out to me that the Tully Monster deserves its special status because it is paleontologically important, prized by collectors, and unique to the Prairie State.

The concretions are most often oval or lozenge-shaped. When they are lightly rapped on the edge with a rock hammer, they crack open along the plane of the impression. It is a wonderful moment when a beautifully preserved section of an ornate seed-fern frond receives the light of the sun for the first time in more than 300,000 millennia.

The second and much less attractive fossil treasure is found in the form of coal balls, dense and often irregular lumps of fossilized peat that did not turn to coal even though they were situated in the coal vein. To the miner and the collector alike, the coal ball has no redeeming value; but paleobotanists will gladly haul them by the burlap-bagful back to their laboratories. When the coal balls are sliced into thin sections with a specially designed power saw, a host of plant forms, including internal structures, can be found. This technique also involves the use of sheets of acetate that extract the patterns of the tissues from the thin sections. When the acetate "peels" are studied in sequence, they reveal the three-dimensional structures—and even the cellular makeup—of the plants entombed in the coal balls.

I refrain from recommending good collection sites in the strip-mined land in this area. If you're dead set on concretion-hunting, ask the local folks in Braidwood, Coal City, and surrounding communities. Virtually all of this land is private property where no collecting is allowed without the owners' express permission. In addition, most of the good sites have already been scoured by so many people over the decades that you might end up with nothing more than your patience badly frayed.

One stop I heartily recommend is Goose Lake Prairie State Natural Area, located about 4 miles west of where the Des Plaines and Kankakee Rivers join to form the Illinois River. At the Goose Lake nature center you will find a fine collection of Mazon Creek flora specimens on display. It may not be as extensive a showplace of fossils as Chicago's Field Museum, but it is informative. The impressions you'll see here may offer you some suggestion of the complexity and abundance of Coal Age vegetation. The Pennsylvanian period's steamy tropical climate was, to put it mildly, very encouraging to the uninhibited growth of plants blessed with an abundant water supply.

During this slice of the geologic time scale, our continent straddled the equator. And at this point, or not long thereafter, North America docked with the earth's other great landmasses to form the supercontinent Pangaea. For this reason, the remains of Coal Age forests are almost identical to Illinois's over a huge range: in Kansas, Pennsylvania, Rhode Island, Great Britain, North Africa, the Czech Republic, and in many other places foreign and domestic. The Pennsylvanian was a unique chapter in earth history. Vast lowland forests of primitive, nonflowering plants such as *Lepidodendron, Sigillaria, Pecopteris, Calamites,* and *Cordaites* flourished and died as the shoreline of the continental sea repeatedly oscillated back and forth. What caused these many cycles of flooding and reemergence? Perhaps the controlling factor was the way new land was formed by the construction of river deltas that invaded marine environments and then subsided. Perhaps the sea level varied in accordance with the accretion or melting of glaciers closer to the poles. Whatever the mechanism, no other period has rock formations that vary so much in type within a larger context of stubborn repetition.

This repetition of rock types ensured that the vast oval accumulation of sediments known as the Illinois Basin would contain more than forty distinct coal beds. Of these, none has been more exploited here near the basin's northern rim than the Colchester coal. (Old-timers still call it the No. 2 coal. This numbering system is not consistent from state to state, and geologists are gradually abandoning it.) In the counties bordering the upper Illinois River, the Colchester lies close to the surface; hence, twentieth-century coal extraction here has largely been through strip mining. Since the strip mining in many locales was finished before the enactment of environmental regulations, much of the land looks like a World War I battlefield about ten years after the Armistice. From the interstate, you see a lumpy and jumbled terrain, with lakes in many of the abandoned pits and scrubby vegetation covering the high ground.

When you take the stretch of Interstate 55 near Braidwood, or when you wander about Will and Grundy Counties and then proceed westward to La Salle, Putnam, Marshall, and Woodford Counties, you will find that many of the towns have other distinctive landmarks: conical hills, often deeply gullied, that almost resemble cinder cones. On more than one occasion I have tried to persuade the unsuspecting that this constitutes the "Illinois Volcanic Field." In fact, these hills are emblems of the region's now-defunct underground coal mines. In the nineteenth century, the only technologically feasible method of mining was the underground variety. In this area, mining concerns employed what was called the longwall method. A vertical shaft was sunk to the coal seam, and miners excavated outward in a full

137

A tree-covered spoil pile from a defunct underground coal mine in Wenona.

The Marshall County town of Wenona has included a spoil pile in its civic image.

circle from a central column of undisturbed rock. Unlike the contrasting room-and-pillar method, where much of the coal and surrounding beds must be kept untouched to provide support against cave-ins, longwall mines were intentionally designed so that almost all of the coal could be removed. This was done by placing timbers and waste rock where it would mitigate the extent of roof collapse in mined-out sections. As long as the roof remained intact over the current work face and haulage ways back to the shaft, subsidence elsewhere was not a great concern.

It would perhaps be impossible to find any coal miner at any time in history who has had a totally stress-free workplace environment, but conditions in the longwall mines must have been particularly hellish. During the peak of the longwall production—the late nineteenth and early twentieth centuries—miners worked at the coal face with nothing more technologically advanced than a pick and shovel, and they usually wielded these while kneeling in a dark, dank, sooty space with a ceiling lower than a standing man's shoulders.

The conical spoil piles that stand as the last testament to dead mines are composed of the waste materials separated from the coal by the aboveground processing plants. These heaps (also known as gob piles and slack piles) contain fossil-bearing concretions along with much fouler stuff: toxic mud, gypsum dust, sulfuric acid, and combustible materials that sometimes ignite spontaneously and burn for extended periods. Some of the piles—for example, the one along Interstate 39 at Minonk—support no vegetation, erode rapidly, and are an arresting brick-red color. Others, such as the one in the little town of Wenona, also off I-39, have managed to develop a cover of tree growth.

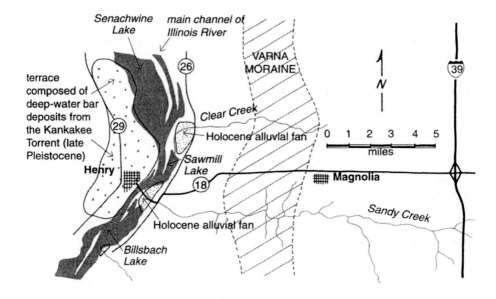

The Illinois River town of Henry and its environs.

— 18 —
Across a Lazy River
HENRY AND ENVIRONS
Marshall County

The valley spirit is not dead;
They say it is the mystic female.
Her gateway is, they further say,
The base of earth and heaven.
—Lao Tzu, *Tao Te Ching*

One of the most fascinating things about the watercourses of the Prairie State is that they tend to have different segments of different ages. From the expansive Mississippi and Ohio to the relatively modest Sangamon and Apple, Prairie State rivers have surprisingly complicated histories. And this is particularly true of the Illinois River.

As noted in essay 7, the upper section of the Illinois—from its head in Grundy County to the Big Bend at Hennepin—is its most youthful part, carved by the waters of the Kankakee Torrent. In the vicinity of Hennepin, however, the Illinois commandeered the channel of no less a river than the ancient Mississippi, which for thousands of years had flowed from the vicinity of Fulton in Whiteside County to the current Big Bend area, then east of Pekin and down to the Illinois River's current mouth above St. Louis. During the Pleistocene, the Mississippi was only grudgingly dislodged from its old course; but finally, approximately 20,000 years ago, the Wisconsinan Shelbyville ice sheet succeeded in doing what the Illinoian glaciers had not: it permanently drove the great stream out of its well-established valley, far west to its current state-border position.

The massive volume of meltwater from glaciers to the north and east created a new outlet to the now-distant Mississippi. The Illinois, the main conduit in this region, occupied the old valley here; and an almost inconceivable amount of water surged past and over the sites of modern-day Henry and Peoria at periods of maximum discharge in the late Wisconsinan stage. But with the disappearance of the glaciers and the change in Lake Michigan's drainage to the north and east, the Illinois River became a much tamer waterway.

The view eastward from the terrace in downtown Henry. The wooded low area on the far side of the Illinois River is the Sandy Creek alluvial fan. The fan has forced the river to swing around it, through a constricted passage.

The motorist who takes Route 18 westward from the Putnam County town of Magnolia to the river settlement of Henry can easily find many interesting geology lessons along the way. The road first rises over the Wisconsinan Varna moraine, then promptly plunges down the high eastern bluff of the Illinois. The tributaries here have done an impressive job of cutting downward over a short horizontal distance. Because their gradient is so steep, they can produce a high water velocity that carries sediments down to the master river very effectively.

Route 18 quickly reaches the floodplain of the Illinois and makes a northwestward swing as it approaches the river bridge. This low ground is a fan of sand and silt—stratigraphers consider it part of the Holocene Cahokia alluvium. This large deposit of sediment, a post–Ice Age feature, has been dumped into the river by Sandy Creek, the feeder stream directly across from Henry. But while tributaries such as this one have the capability to deliver the alluvium, the much wider waters of the Illinois lack the will to remove it. In this stretch, the Illinois is one of the world's flattest rivers— its gradient is very low. It does not have the energy necessary to sweep away the fans' sand and silt to keep the channel clear. Instead, it winds feebly around the obstructions and backs up to form the floodplain lakes and wetlands that are characteristic of the middle Illinois. Geomorphologists call this sort of river—one that receives more sediment than it can remove— an aggrading stream.

142

The drive across the bridge provides a good look at two of the floodplain lakes (one upstream, one down) that have been formed by fan-building. The town of Henry occupies an excellent position on a high, broad bench on the river's western side. Can you figure out why this natural, high-level platform attracted human settlement? It has much to do with the fact that it offers both safety from latter-day floods and easy access to the waterfront. Recent analysis suggests that this surface is composed of deep-water bar deposits brought here by the Kankakee Torrent about 15,500 years ago. The highest point of the terrace is well over 50 feet above the river's current level—and that gives you a good idea of how high the torrent's turbulent waters were. The Great Flood of 1993, which caused the middle Illinois to flow backward and almost change its course in the Peoria area, was a mere trickle in comparison.

Henry has two excellent vistas: one down below, at the waterfront park, and one up above, from the bar-deposit bench. Below the river's bed and the Sandy Creek fan rests a huge deposit of much older Ice Age outwash. Of Pre-Illinoian age, it is known as the Sankoty sand. It may be invisible at the surface, but it plays an important role in the lives of thousands of middle Illinois Valley residents, because it is the region's main aquifer, or groundwater-bearing layer. The meltwaters that deposited the Sankoty sand far back in the Pleistocene are long gone; but their buried outwash still continues to influence the modern world.

The eastern bluff of the Illinois River just north of Henry. Note the large quarry face. The sloping cut face of this exposure suggests that the material is sand or gravel outwash from the Kankakee Torrent and not loess, which forms vertical cuts.

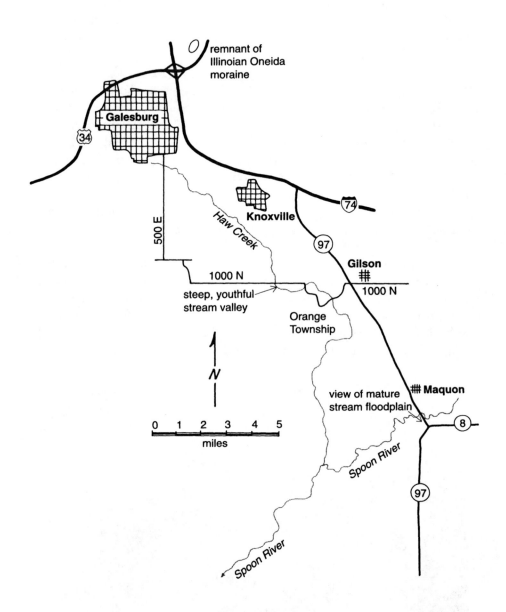

Galesburg and vicinity.

― 19 ―
A Landscape Throws off
Its Glacial Burden
THE GALESBURG PLAIN
Knox County

This earth, like the body of an animal, is wasted at the same time that it is repaired. It has a state of growth and augmentation; it has another state, which is that of diminution and decay.
—James Hutton, *Theory of the Earth,* 1795

In the past 1.6 million years, almost nine-tenths of the Prairie State has felt the weight of the glaciers. The areas discussed in many of the preceding essays bore the brunt of the last glaciation of the Pleistocene, the Wisconsinan (75,000 to 10,000 years ago). Now we consider a portion of Illinois that was last covered by the older and much more far-reaching ice sheets of the Illinoian stage (about 300,000 to 125,000 years ago).

In contrast to the johnny-come-lately Wisconsinan terrain, where the Ice Age features look as though they were formed last Saturday afternoon, Illinoian landscapes have had roughly ten times as long to suffer Hutton's state of diminution and decay. If you travel to the flat stretches of farm country in McLean County, for instance, you'll find a surface still so thoroughly plastered with Wisconsinan glacial till that it is really featureless. With the exception of farmers' straight ditches and drainage tiles, there are few if any streams. Here in Knox County, however, the tributaries of the major rivers have had a longer time to work their way headward and reestablish a drainage net. The physiographic section that contains Knox County and much of west-central Illinois is the Galesburg Plain. It is one of the best places to see a landscape in the process of removing its glacial past. Its streams are hard at work carrying away the drift that the Illinoian ice sheet left in abundance.

Geomorphologists, the specialists who study landscapes and their development, have always had a particular interest in streams, and for good reason, since streams are often the driving element in landscape evolution. One of the most influential of all geomorphologists, the American William Morris Davis (1850–1934), devised a simple classification system for the stages of stream development that is still in use today. Part of its appeal lies

The steep and narrow Haw Creek valley, along 1000 North in Knox County's Orange Township.

in the directness of its metaphor: drainage systems are compared to living organisms at the different stages of their life cycle. In the first stage, youth, the land is made up of large tracts of still undrained land separated by tributaries that have V-shaped cross sections and do not meander a great deal. Downcutting and valley-deepening is the order of the day. In maturity, the tributaries have extended themselves farther. Almost all of the land is well drained, and downcutting is reduced in favor of the streams sidecutting their valleys. Most of the terrain consists of hillsides rather than broad, undissected upland. And at the final stage, old age, streams meander widely across broad floodplains, and the relief of most of the terrain is subdued. The process of erosion has practically come to a standstill—until renewed crustal uplift or some other major change starts the cycle anew.

If the streamless Wisconsinan till plain stands at the very beginning of this cycle, then the Galesburg Plain represents the youthful stage, or, in some cases, maturity. Start at the intersection of Routes 8 and 97, just south of Maquon. If you head north on 97, you'll first cross the valley of Illinois's number-one literary river, the Spoon, far above its debouchment into the Illinois River across from Havana. It is the major stream in this area, and as is often the case, it has reached a later stage in the cycle than its tributaries. The presence of meanders and a broad floodplain points to its having attained maturity. But in heading north toward Knoxville, especially on the unpaved side roads, you will find long stretches of horizontal till plain suddenly punctuated by steep drops into the V-shaped tributary valleys described above. My favorite example of this is in Orange Township, where 1000 North plunges down to Haw Creek in classic Galesburg Plain style. It is here that erosion is keenest: the modest tributary can be compared to a circular saw blade that is cutting downward and forward (that is, upstream) at the same time. It has much virgin upland surface to devour—and in the centuries to come, it will.

To the person accustomed to ridge-covered northeastern Illinois, the flat upland surface of the Galesburg Plain is conspicuous in the absence of end moraines. In general, Illinoian landscapes do not have many impressive features of that sort. (For a particularly nice exception, see the description of the Buffalo Hart moraine at the end of essay 25.) Illinoian moraines have had longer to erode than their newer Wisconsinan counterparts. But the Illinoian ice sheet does not seem to have built many moraines in the first place. Perhaps the climatic conditions of the Illinoian stage promoted rapid glacial advance followed by rapid glacial downwasting. This would have discouraged extensive end-moraine formation, which depends on a long, steady-state period in which an ice sheet melts at the same rate it moves forward.

In a spot north of Knoxville, though, you can track down a subtle example of an Illinoian end moraine. It can be seen from Route 34 just east of the junction with Interstate 74. As Route 34 curves northeastward and parallels the railroad tracks, look on the north side of the road. You should spot a low mound—only about 20 feet in relief—with trees and a farmhouse on top. This is a detached section of the Oneida moraine. Off and on, it runs some 15 miles northeastward from this, its southern tip.

The low rise of the Illinoian Oneida moraine is highlighted by trees and farm buildings. Route 34 and railroad tracks run in front of the moraine.

147

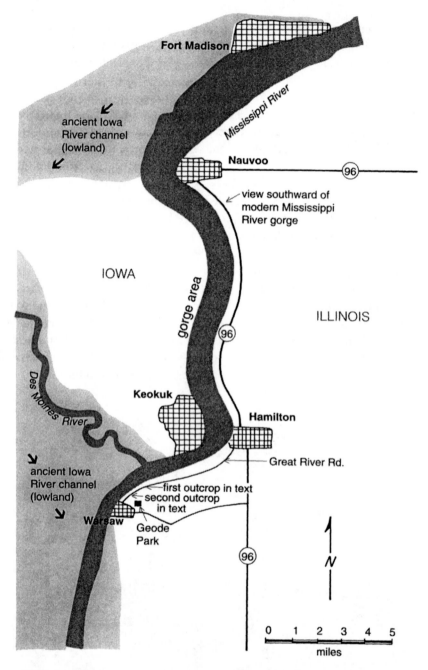

Mississippi River country and gorge from Nauvoo to Warsaw. The light gray shaded region is a lowland created by the ancient Iowa River.

Homage to the Mississippian
NAUVOO TO WARSAW
Hancock County

*At the time that any stratum was being formed, all the matter resting
upon it was fluid, and, therefore, at the time when the lowest stratum
was being formed, none of the upper strata existed.*

—Nicolaus Steno, 1638–1687,
De solidarum intra solidum

Most of Illinois seems to be one vast demonstration of Steno's simple
but all-important observation that in an exposure of flat-lying sedimentary
strata, the oldest beds are at the bottom, and the youngest beds are at the
top. There are a few places in the world where the powerful forces of
mountain-building have flipped strata completely upside down; but we need
not worry about that prank of nature here in our continent's interior.

This essay is addressed first and foremost to rock hounds who delight
in the mysteries to be unraveled at any good roadcut or cliff. To the geologist or amateur enthusiast, the different components, structures, and textures
of a rock formation can be as illuminating and aesthetically pleasing as an
orchid's complex floral anatomy is to a botanist. The Mississippi River country
from Nauvoo to Warsaw is a superb place to scrutinize four formations that
were laid down in the early Mississippian period. (Don't be confused by
these two separate terms: Mississippi always refers to the river; Mississippian
to the Paleozoic period, which lasted from 360 to 320 million years ago.)
It is also an ideal locale to use Steno's rule to travel upsection—that is,
forward in time, from the lower, older beds to the upper, younger ones.
Mississippian rocks outcrop along much of the lower two-thirds of Illinois's
western boundary, but it would be difficult to imagine a more beautiful
stretch of the river to explore.

Traveling southward on Route 96 from the handsome, restored Mor-
mon settlement of Nauvoo, you'll quickly notice that the river is confined
to a narrow passage between rock bluffs. This gorge was cut by a different
stream early in the Pleistocene epoch, when the Mississippi was still flowing
far east of here, in a course that took it to the site of modern St. Louis by
way of central Illinois. In those days, the river that was draining this area

was the ancient Iowa River. Originally, it took another route between Nauvoo and Warsaw—through a valley ten times as wide as this one, a few miles to the west. But that much roomier route was blocked by the Pre-Illinoian glaciers and their outwash, and the stream had to carve this new, confining channel through the highly resistant Keokuk limestone. Still later in the Pleistocene, during its concluding Wisconsinan stage, the Mississippi was pushed westward by the Shelbyville ice sheet, where it appropriated the ancient Iowa's valley, including this youngest segment.

For many years, this segment of the Mississippi contained the Lower Rapids, a barrier to navigation that was circumvented by the construction of a dam between Keokuk, Iowa, and Hamilton, Illinois. As you approach Hamilton, watch for the signs that keep you on the Great River Road south of the downtown area. (Route 96 here takes a separate course heading inland.) Mark your mileage when you reach the western intersection of Routes 96 and 136. At approximately 2.8 miles down the Great River Road from that junction, you'll come to a large roadcut on the eastern side of the road. This is the first of three sites best inspected on foot, so park well over on the shoulder and carefully cross the road. Warning: traffic here is heavy and fast. Keep one eye on the geology and one on the road.

What stands before you is an excellent exposure of the upper part of the Mississippian Keokuk limestone, the same rock that floors the Lower Rapids gorge. Your inspection will confirm that this is mainly carbonate rock—primarily dense gray limestone, with a little dolomite at the top, and some dark-toned calcareous shale layers. The limestone is the type known as biocalcarenite—it is composed of sand-size bits of skeletal remains. In this case, it also contains entire fossil crinoids (animals nicknamed "sea lilies") and lots of chert nodules. The chert is an extremely hard material, and it no doubt accounts for much of the Keokuk's brave stand against the river's awesome erosive power. A form of quartz, chert probably had its source in the hard parts of sponges and other silica-secreting marine organisms.

This outcrop is just one of hundreds of Mississippian exposures that can be easily studied in the Midwest. What conditions were necessary for the creation of so much limestone extending over thousands of square miles? The first thing to keep in mind is that the Mississippian in Illinois was predominantly a marine period, when a shallow saltwater sea covered the Illinois Basin and surrounding areas. Southward, that sea connected with the open ocean; dry land lay to the north. During this period, rivers poured out massive quantities of sediments from various mouths along the coast (and not, as was thought previously, from one predominant river that built a large bird's-foot delta like the one at the mouth of the modern Missis-

The Keokuk limestone as it appears in the northernmost roadcut described in this essay.

sippi). These rivers discharged most of their load northeast of here, but also, to some extent, to the north and northwest. As the sediments accumulated on the seafloor, and as the world's sea level changed, the basin continued to subside, and many miles of the coastline continued to shift back and forth, alternately northward and southward. The Illinois Basin was well on its way to becoming the vast repository of sedimentary rocks that it is today.

The types of marine rocks you find at an exposure like this often indicate, in a rough sense at least, how far away the delta and the coastline were at the time of deposition. Limestone and dolomite, precipitated locally from carbonates in the water and from the remains of shelled organisms, imply in this case that the land was quite distant. Shale, derived from mud, suggests a somewhat closer shoreline; but sandstone—composed of larger, heavier particles that cannot be transported as far as mud—is the best sign of the land's close proximity. Why was the shoreline moving about so restlessly in the first place? For one thing, the sea level might have changed in response to varying amounts of sediment being discharged by the rivers to the north. And the process of plate tectonics—in which landmasses moved into areas previously occupied by deep ocean basins—might have caused the water height to vary. There are other plausible explanations. We'll probably never be absolutely certain.

A look at the second Mississippian outcrop mentioned in the text. The lower half is shale and limestone of the Warsaw formation. Above it lies a fairly narrow bed of massive Sonora formation sandstone, which projects beyond the Warsaw beds. The uppermost layer belongs to the St. Louis formation.

When you've had ample time to look over this first Mississippian roadcut, continue southward another 1.1 miles. The road takes you uphill—which, because the strata are basically horizontal, also means you're going upsection, toward younger units. When you see the large outcrop on the eastern side, carefully pull off the road. Again you will have to take special care in crossing the busy highway and walking underneath potentially loose rocks. This exposure reveals the formations that lie just above the Keokuk limestone. Starting at the bottom is a group of strata that form a sloping, apronlike surface made up of both limestone and dark gray shale beds. This is the uppermost section of the Warsaw formation. It sits directly atop the Keokuk, which here is well out of sight. Above the Warsaw, about halfway up the cut, is a foot-and-a-half layer of sandstone weathered to a brownish tint. This is the Sonora formation, which in other locales can be 20 feet thick or more. Above that stands the St. Louis limestone, the youngest Mississippian formation described in this essay. It is termed a lithographic limestone because it is the kind of dense, fine-grained rock that lithographers used in times past for their engraving plates.

Another interesting thing about the St. Louis formation is that it contains sinkholes and evaporite deposits that point to its being exposed to the open air during a retreat of the Mississippian sea. If you look at the base

A large streamcut of the Warsaw formation at Geode Park. The Warsaw tends to form sloping, easily eroded cut faces because it is predominantly shale.

of this formation here, you'll see its beds are distorted and broken up into breccia, or angular blocks cemented together. Decades ago, this breccia was thought to be the result of surf action working on the exposed limestone, or from some sort of large-scale deformation in the area. More recently, experts at the Illinois State Geological Survey have theorized that it was caused by intense swelling and cracking when the evaporite mineral anhydrite was precipitated and subsequently transformed into gypsum. In the end, the gypsum was dissolved away.

To see the final outcrop on our list, you must slide downsection a bit, to another portion of the Warsaw formation. It is exposed magnificently in the town that bears its name, at the alluringly named Geode Park. The park entrance is on Main Street, just a little east of the town center. (As you approach it, pay your respects to the ghost of Warsaw's most illustrious nineteenth-century resident, Amos Worthen. At one point, Worthen ran a shop in town, but he is much better known today as the man whose pioneering fieldwork and writings dramatically advanced the knowledge of Illinois's geology.) When you spot the Geode Park sign, proceed downhill, past the red barn, the barking dogs, and the picnic shelter. At the bottom is a tall, creek-cut exposure of the Warsaw formation's lower part. As the park name suggests, this group of strata is the source of top-notch geodes, and has given this area the reputation as the Geode Center of the Known Universe. This site has been so picked over by collectors that your luck may not be very good, even if you concentrate your efforts on the other local attraction, the corkscrew–shaped fossil bryozoan, *Archimedes*. This creature, a sedentary, colony-forming invertebrate, is (or in this case, was) so abundant in the Warsaw formation that early stratigraphers called the rock the *Archimedes* limestone.

Geodes have such broad appeal, even to people not interested in geology or rock collecting, that it's surprising their origin is still something of a mystery. Geologists agree that they are secondary structures developed in small cavities in the host rock; but how the cavities come to be, and how the distinctly different concentric sections of geodes form, is the stuff of complicated theories. On the outside surface, geodes have a thin clay coating. Inside the coating is a lovely zone of chalcedony, a form of impure quartz often called agate. Chalcedony is a substance without recognizable crystal faces; often it is steely bluish gray, but it can be other colors. Inside the chalcedony rest inward-pointing crystals of quartz, calcite, or, less frequently, some other mineral. (Warsaw geodes sometimes contain even petroleum or bitumen.) One leading theory accounts for the geodes' internal pattern by proposing that as the initial hole is filled with seawater and an outer barrier of gelatinous silica it becomes an osmotic cell that absorbs fresh water without releasing it again. Eventually, the geode dries out, cracks, and lets in other minerals, which form the innermost crystals.

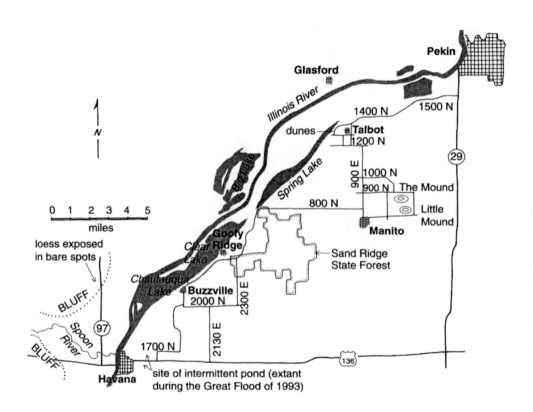

The Illinois River and environs from Pekin to Havana.

— 21 —
The Havana Lowland
PEKIN TO HAVANA
Tazewell and Mason Counties

Some landscapes, one learns, refuse history; some efface it so
completely it is never found; in others the thronging memories
of the past subdue the living.
                    —Loren Eiseley, *The Invisible Pyramid*

If there is a site in Illinois where the landscape of the past subdues the
living, it is the broad, wedge-shaped bottomland stretching between Pekin,
northern Logan County, and Beardstown. It is a place where a valley much
larger than its present river needs holds a superabundance of clues about
great floods and vanished streams.

If you had visited the Havana Lowland early in the Pleistocene, about
one million years ago, you would not have found the Illinois River flowing,
as it does today, across a plain choked with glacial outwash. You would have
come upon the ancient Mississippi, which in those days came this far east
before swinging back southwestward to the St. Louis area. Near the modern
town of Delavan, this great river was joined by a stream that may have been
even greater: the Mahomet River. The Mahomet—known by geologists in
states to the east as the Teays—flowed all the way from West Virginia, through
Ohio, Indiana, and east-central Illinois to this grand junction, with such
major tributaries as the Middletown and Danvers Rivers. Together, these
streams swung back and forth through the relatively unresisting Mississippian and Pennsylvanian strata of the area to produce a huge network of
bedrock valleys.

There is evidence that Pre-Illinoian glaciers had covered the region by
the middle of the Pleistocene, but the Havana Lowland and its stream courses
remained more or less intact until the arrival of the most extensive ice sheet,
the Illinoian. The tributary valleys disappeared under a blanket of glacial
drift, and the lowland was partially filled in. By the end of the Illinoian stage,
a newer system of tributaries, the one we know today, apparently had begun
to establish itself.

The ice sheet of the final Pleistocene stage, the Wisconsinan, did not
reach the lowland. Ironically, its influence on the region was more profound

The ancient drainage pattern of the Havana Lowland region. The sites of modern cities are shown, along with the approximate extent of the modern Havana Lowland (gray).

than anything preceding it. About 20,000 years ago, the Wisconsinan glacier's farthest advance, the Shelbyville, finally dislodged the ancient Mississippi from its accustomed valley and repositioned it in today's course far to the west. Then, about 1,000 years later, the Bloomington glacial advance reached the Peoria area and built both a damlike moraine and a large outwash deposit south of the moraine. Later still, dramatic episodes of rushing meltwater flooded this lowland. One of these floods, triggered some 15,500 years ago by morainal-dam breaches upstream, was the mighty Kankakee Torrent. The force of the torrent in this area was so great that the lowland's preexisting Pennsylvanian rocks and old channel sediments were often torn from their resting places, jostled about in the surge, and eventually deposited downstream. In quarries south of Pekin, one can sort through the waste piles and find two variations of a unique geological phenomenon, armored till balls and armored mud balls. These are lumps of till or silt that were bounced along the river bottom by the strong Kankakee Torrent currents and coated with smaller pebbles of many different origins. I have one specimen that is the size of a basketball, weighs about thirty pounds, and is coated with everything from Precambrian Baraboo quartzite from Wisconsin to Silurian dolomite from the Lake Michigan area. And it also contains local materials, including chunks of Pennsylvanian coal.

In the postglacial (or perhaps interglacial) setting of the Holocene, the Illinois River has been much more well behaved—to the point of being

Massive crossbedding patterns in Kankakee Torrent sand and gravel deposits at a quarry southwest of Pekin. Because the sediments here are coarse-textured and unconsolidated, slumping and gravel slides are common—as the large talus fans at the base of the cut demonstrate.

downright lethargic. Geologists call it an aggrading river, one that is building up its bed rather than cutting down through it because it receives more sediments than it can carry away. In the neighborhood of Havana, it has an extremely flat gradient of only 3 inches drop per horizontal mile. No wonder it backs up so readily when the Mississippi floods downstream.

Begin your exploration of the Havana Lowland at its northeastern tip; then work your way southwestward. South of Pekin, the best bet is to turn off Route 29 and head westward on the county roads—only some of which are paved. When Illinois State Geological Survey educators do a field trip for their loyal following of Pekinese, they stop at two beautifully illustrative sites. The first, an excellent roadcut through a sand dune, is located west of the small town of Talbot, just north of Spring Lake State Wildlife Refuge, and south of and across the Illinois River from the ancient Glasford meteorite-impact site. Hundreds of millions of years of erosion have rendered the Glasford feature, about 2.5 miles in diameter, indistiguishable at the surface.

By carefully examining the sand grains at both ends of the dune roadcut, you can determine what the prevailing wind direction was when this landform was being built up. The larger grains have fallen at the head of the dune; the smaller ones, farther downwind.

A dune of Parkland sand, southwest of Pekin. The prevailing winds that formed the dune moved from left foreground to right background.

This sand dune and the ones nearby sit atop the highest of three Havana Lowland features that were once described as valley terraces. Recent field research by geologist Edwin Hajic suggests that in some cases, at least, these "terraces" may not have been created by various floods in a relatively orderly progression over a long span of time, but suddenly, by one or more spates of maximum torrent flow.

The level here is the so-called Manito terrace. It stands between 480 and 500 feet above sea level. That's about 100 feet lower than Lake Michigan, but high by this area's standards. In some places, the Manito terrace and its two lower counterparts, the Havana and the Bath, are not continuous surfaces but detached segments. One of the most dramatic of these is The Mound, a hill that stands between the Breedlove and Hickory Grove ditches, on 900 North a little above the Mason County line. It is an expression of the Manito surface. One-half mile south of The Mound stands—you guessed it—Little Mound, a fragment of the next level down, the Havana terrace (465 to 480 feet above sea level). As for the lowest terrace, the Bath, you're driving on it when you're on the relatively low ground on either side of The Mound.

No drive in this area would be complete without a transit of Sand Ridge State Forest, a large plantation of pine trees that thrive in the lowland's

160

sandy soil. Also, a swing through the quintessential Illinois River settlement of Goofy Ridge should be obligatory. The folks in towns such as this are often called River Rats by uplanders, and by themselves, too. Despite the harrowing events of recent years, they continue to occupy their floodplain real estate with a sense of unruffled calm that makes the geologist, all too familiar with the power of rising rivers, wince. Periodically their cars, dogs, and bedroom additions end up in another county, transported there by the foaming brown water of flood.

Farther south, you can catch another glimpse of Manito terrace sand dunes by taking Route 97 southeastward out of Havana. In this same locale, a large intermittent pond formed during the Great Flood of 1993. In that year, the water table in the saturated glacial outwash was so high that it literally saw the light of day. The Illinois Department of Transportation spent much of that summer raising the level of the road, but the railroad tracks and several businesses were flooded.

Before you leave the Havana area, retrace your track on Route 97 and cross the bridge over the Illinois. Continue across the expansive fields (the site of one of the state's largest farms) and head up the bluff on Route 136. The 3- to 4-mile-wide floodplain, and the much broader lowland that

Participants in an Illinois State Geological Survey field trip walk down a road in front of The Mound, a remnant of the high Manito terrace. In the foreground, the road runs across the much lower Bath terrace.

161

contains it, make the Illinois a classic example of an underfit stream—one that is much too small for the valley it occupies. Here you can easily imagine an earlier time, when the ancient Mississippi and its attendant rivers snaked through this great expanse. Continuing along on Route 136, you will see one of the most abrupt changes in terrain anywhere in the state, where the Havana Lowland suddenly gives way to the deeply dissected upland of the Galesburg Plain. An examination of the lowland's western bluffs, especially on Route 97 north of the mouth of the Spoon River, is a rewarding side trip. You'll see thick deposits of loess (pronounced *luss*), a tan or buff-colored windblown silt, mantling the slopes. The deepest loess deposits are usually on the eastern side of river valleys, because of the prevailing westerly winds. Here, there is no lack of it on this western side.

This very fine material, found over much of the state, was originally deposited on river floodplains in the Pleistocene. In the warmer months, the meltwaters would be at their highest annual level; but in winter, the ice sheet melted less, and the streams accordingly shrank. That left the silt deposits exposed to the air. Because silt is lighter than sand, it is easily carried farther by the wind. As you can see in roadcuts and bald spots on the farmland in this vicinity, the wind had no problem hefting large quantities of loess to a great height—in places, well over 100 feet above the river's surface.

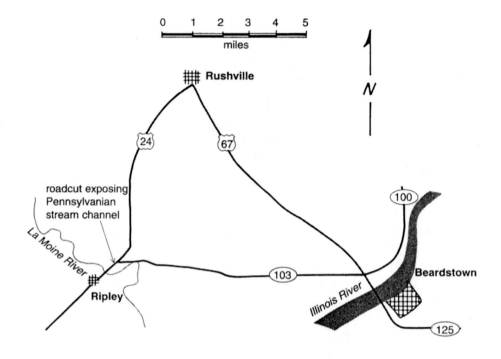

The Pennsylvanian-Mississippian stream-channel roadcut is just north of
the U.S. 24 bridge over the La Moine River north of Ripley.

Reconstructing a Coal Age St

NORTH OF RIPLEY
Schuyler County

Time is but the stream I go a-fishing in.
—Henry David Thoreau, *Walden*

Flowing streams have long served as an allegory for the passage of time. They embody its separate attributes of impermanence and seamless continuity. Near the small town of Ripley, on the border of Brown and Schuyler Counties, the waters of the La Moine River make their way eastward toward the broad valley of the Illinois. They also mark the place where the passage of time is admirably displayed in the record of the rocks. You can take a close look at this pleasant stream on the Route 24 bridge just north of Ripley. Unless the water is high, you should be able to see sand and silt deposits exposed on its bank. These sediments, probably derived from glacial drift of the Galesburg Plain some miles upstream, may eventually continue their slow descent toward the Gulf of Mexico.

It's a fairly safe bet that the La Moine will continue its work of erosion, deposition, and transportation for centuries to come. Imagine, however, that the world's oceans suddenly rose so high that salt water covered central Illinois. The La Moine's channel would be entombed in new marine sediments, and both they and the riverbed's sand and silt would eventually be turned to stone. In effect, this stream would become a well-preserved fossil feature awaiting some future geologist who might chance upon it and decipher its ancient role.

If this scenario sounds farfetched, walk along Route 24 north of the bridge and take a good look at the large roadcut on either side of the roadway. Believe it or not, you are staring at just such a fossil streambed, revealed in cross section. You are also facing rocks of two different geologic periods, the Mississippian and the Pennsylvanian. Geologists in many parts of the globe lump these two periods into one, which they call the Carboniferous. In the United States, however—and particularly in Illinois—it makes good sense to keep them separate, because they represent spans of time with conditions that were quite distinct from one another.

The heart of the Pennsylvanian stream-channel fill just north of Ripley. The strata bow downward into the older streambed, which was carved into the underlying Mississippian Salem limestone.

The line between the Mississippian and Pennsylvanian beds here is clearly exposed at the northern end of the outcrop. Near road level you will see strata formed of a dense carbonate rock that weathers a pinkish white. This is the Middle Mississippian Salem limestone, the same formation that contains Indiana's famous Bedford limestone, used extensively as architectural cladding. At this location, the carbonate rock is most precisely described as dolomite. It is not so much calcium carbonate, as in true limestone, as calcium-and-magnesium carbonate.

Just above the Salem beds is a series of other sedimentary rock types. Heading upward, you find sandstone, layers of gray and black shale that erode easily and form a large hollow on the cut face, more sandstone that projects outward boldly and has prominent crossbedding patterns, and siltstone near the top. All these layers are part of the Early Pennsylvanian Tradewater formation (previously known in Illinois as the Abbott formation). Taken as a whole, this outcrop is by no means an example of continuous deposition: at the boundary of the dolomite and the sandstone there is a time gap of at least 20 million years. Originally, there were other, younger

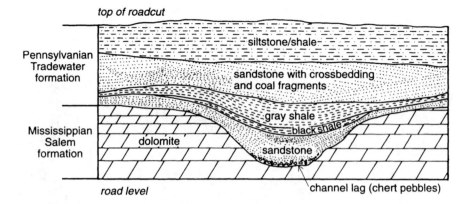

Pennsylvanian
Tradewater
formation

siltstone/shale

sandstone with crossbedding
and coal fragments

gray shale

black shale

Mississippian
Salem
formation

dolomite

sandstone

road level

channel lag (chert pebbles)

Cross section of the Pennsylvanian-Mississippian
stream-channel roadcut. —Illinois State Geological Survey

Mississippian rock layers here—maybe hundreds of feet worth. At the close of the Mississippian period, however, the sea that had covered this area withdrew and left the strata exposed to erosion. Everything down to the Salem formation was obliterated. This sort of an erosional interruption in the sequence of deposition is termed a disconformity.

Later, at some point in the Early Pennsylvanian, a small stream about the size of the nearby La Moine established itself on the remaining Mississippian rock you see here. Its channel can be traced by following the slight downward curve of the high, crossbedded sandstone that is most obvious at the outcrop's center. Below it, the shale and lower sandstone mark the narrower, bottom part of the streambed. If you're lucky, you may find some thin patches of conglomerate made up of chert gravel directly above the Pennsylvanian–Mississippian disconformity. This gravel represents the channel lag deposit: coarser sediments that bumped along the bottom of the streambed at a slower speed than the finer, suspended material above. The vast preponderance of smaller particle sizes suggests that this creek was anything but a raging torrent. It did not have the energy to carry substantial amounts of coarser materials. This is not surprising, because the landscape of Pennsylvanian–period Illinois was low lying, and most stream gradients were probably quite flat.

A detailed examination of the rocks here can be a delightful experience, but be mindful both of passing traffic and of loose rocks falling from above. On my first visit here, I found two geodes in the talus in the space of five minutes; presumably they came from the Salem beds. They were of poor quality, compared with those found in the Warsaw formation (see essay 20).

The piece of light-toned Pennsylvanian sandstone between the rock hammer and the lens cap contains small strips and chunks of coal—the remains of plant stems that fell into the ancient stream approximately 310 million years ago.

In the talus apron at the foot of the cut face you will probably find many big chunks of the light gray Pennsylvanian sandstone with crumbly black material in them. Can you surmise what this substance is? It is the remains of plants that sank to the stream bottom and eventually turned to coal. Sometimes the details of the plants' anatomy are still partially visible.

These coal traces come from the Pennsylvanian counterparts of the trees lining the La Moine today, but we should not draw too many comparisons. The Coal Age knew no flowers, no butterflies, no birdsong; the breezes of 310 million years ago helped shed the spores of strange whiplike trees more closely related to modern club mosses than to oaks, maples, and other advanced plants dominating our world today. Beneath them grew a clutter of other primitive but successful plant types: seed ferns, true ferns, horsetail relatives, and the predecessors of modern cone-bearing plants. This tropical botanical madhouse had many other tenants, including amphibians, primitive reptiles, and supersized versions of cockroaches and dragonflies. It gives us good reason to be humble: an abundance of exuberant life reduced to a film of coal on a broken slab of rock. It is heartening, though, to realize that two geologic eras later, there is a creature who can reconstruct, at least in part, the story contained in the film of coal. The creature is you. You are fishing very deep in time's waters.

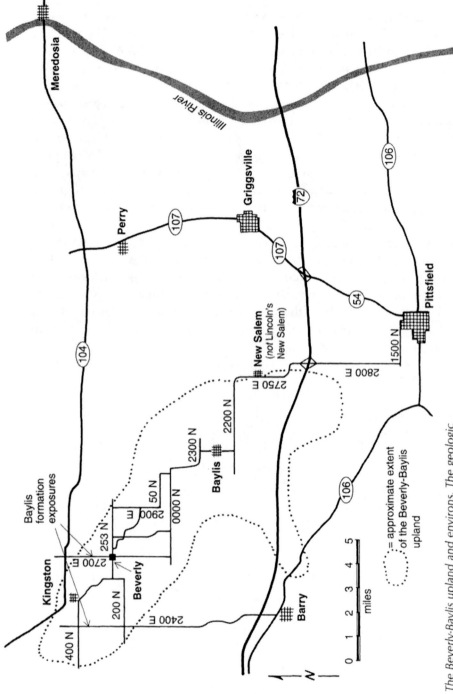

The Beverly-Baylis upland and environs. The geologic tour begins in Beverly, heading north on 2700 East.

— 23 —

A Mystery from the Age of the Dinosaurs
THE BEVERLY-BAYLIS UPLAND
Adams and Pike Counties

So long as there's a hill-ridge somewhere the dreamer
Can place his land of marvels.
                    —W. H. Auden, "Plains"

Since the end of the great span of Precambrian time, there have been three geologic eras: the Paleozoic, the Mesozoic, and the Cenozoic. As all parents of grade-school children know, the second is the hands-down winner of the popularity contest, because it is the Age of Reptiles, when the dinosaurs held sway. Unfortunately, people interested in the geology of Illinois will find that the Mesozoic is the era least well represented. Instead, the record of the rocks offers a splendid array of Paleozoic formations; and the Cenozoic is represented by Tertiary formations at the state's southern tip, and by much more recent and extensive Pleistocene deposits that cover almost nine-tenths of the state.

If you study a geologic map of Illinois, you will search in vain for the first two periods of the Mesozoic era, the Triassic and Jurassic. They left no trace, at least no lasting trace. But the third and final Mesozoic period, the Cretaceous, is represented in two different sections of the Prairie State. The first, and much better known, consists of three formations that are exposed at the state's southern tip. The other is located in rural, west-central Illinois, in an area nicknamed Forgottenia because of its remoteness from seats of political and economic power. This formation was originally thought (and correctly) to be pre–Ice Age, but its place in the geologic time scale was uncertain. Several decades ago, when scientists of the Illinois State Geological Survey analyzed the deposit's mineralogical content, they discovered it was probably from the Upper Cretaceous—about 100 million years old. The real marvel, though, is how these sediments came to be here in the first place.

This west-central deposit happens to be in one of the less frequented spots of the state. It forms a broad, prominent upland trending northwest to southeast through the towns of Beverly and Baylis. Near Baylis it reaches its highest elevation, about 885 feet above sea level and roughly 400 feet

171

above the Mississippi River floodplain, 10 miles to the west. The deposit, known to stratigraphers as the Baylis formation, is up to 100 feet thick; but for the untrained eye it can be tricky to locate. It contains every particle size from clay to cobble, but sand is the primary constituent, with gravel concentrated in the bottom 10 feet. It is almost identical in composition to deposits of proven Cretaceous age some 200 miles to the west, in Iowa, Kansas, and Minnesota. In addition, it lacks the igneous rocks from the Canadian Shield that are so prevalent in Illinois's glacial drift.

Does it seem strange to you that in 100 million years the Baylis formation has not been transformed into stone? The process of lithification—of turning unconsolidated material into rock—is complicated, involving more than the age of a deposit. The Cretaceous sediments of far-southern Illinois are not lithified, either. As a rule, sediments turn to rock only when they are buried deep beneath newer formations and when conditions are suitable for proper compaction and cementation. The Baylis formation probably was never covered by extensive younger deposits.

One theory suggests that the Baylis was deposited in a shallow, nearshore marine environment. Or perhaps it was laid down by a river system some distance from the sea. As you'll observe, the formation's basal section contains large pebbles and cobbles that are well rounded and sometimes glossy. Among them are specimens of red and brown chert and a lovely purplish metamorphic stone of sugary texture that looks for all the world like Wisconsin's Baraboo quartzite. These coarse-grained sediments would seem to have been carried here from a source far north of Illinois. However, the flat upper surface of the underlying bedrock shows that in the Cretaceous period this region was a heavily eroded landscape of low relief, where we would expect the streams to be too sluggish to carry sediment as heavy as gravel. Illinois State Geological Survey scientist John Masters reminded me that such streams can carry gravel as slow-moving channel lag, at the base of their beds. Also, as Masters noted, the gravel might have been transported by periodic floods, which have greater carrying power.

To see this upland, which forms one of Illinois's most unusual geologic features, start at the small settlement of Beverly. Half a mile north of town, on unpaved 2700 East, you can find a thin section of the Baylis formation—in the form of weathered, rusty brown sand and polished pebbles—on a low cut on the eastern side of the dirt road. If you stop and look carefully, you'll see at least the larger bits of gravel poking out at the surface, even though this exposure has been grassed over. (Because of the vegetation cover, a visit in late fall or early spring is your best bet.) As you continue a little bit uphill and upsection, you will come across a deposit of tan silt. This is not the

Sediments of the Cretaceous Baylis formation. Large quartzite and chert pebbles, often polished and rather rounded, sit in a matrix of loose sand. From a roadside exposure along 2700 East, north of Beverly.

Baylis, but younger Pleistocene loess. The loess, a windborne deposit blown here from the Mississippi River valley, has a paler tint and does not contain coarser particles.

More obvious exposures of the Baylis formation can be found farther west, a little beyond Kingston, on 2400 East starting at 0.2 mile south of its junction with 400 North. As you drive along this stretch of gravel road, look at the farm-field gullies along the eastern shoulder. Here the sand and the shiny, well-rounded pebbles can be found in abundance, but they are on private land.

At this point, I suggest you retrace your course to Beverly and take 253 North and the connecting blacktops from there down to Baylis. This is beautiful farm country of rolling terrain and fine vistas. The area has an unmistakable upland feel to it. Keep a close eye on the roadcuts and gulleys, and see if you can tell the difference between the Pleistocene loess and the Baylis formation. Here are some clues: as noted before, the loess is a light tan; it likes to form vertical faces; and it's found on what geologists call "topographic highs"—the tops of hills and rises, rather than deep in valleys and hollows. The Baylis is exposed in low places, has a somewhat darker

Farm-field gullies offer some of the best exposures of the Baylis formation. Photo taken along 2400 East southwest of Kingston.

brown tone, and is too coarse and uncompacted to form vertical faces. (The color is a sign of weathering; beneath the soil zone, the Baylis is white.)

As you enjoy your drive through this high, sparsely populated countryside, ponder one final mystery. How did it happen that material as soft and yielding as this survived as an upland? This land is part of a long, narrow strip extending from Adams County to Calhoun County that is one of Illinois's three driftless areas—areas that the Pleistocene glaciers apparently never covered. Could the lack of glacial action have been a major factor in the upland's survival? And could it be that the upland is the last surviving remnant of a much larger deposit? To give John Masters the final say: "Erosional remnants are rampant in earth history."

Pere Marquette State Park is on Illinois 100, west of Grafton.

— 24 —
The 120-Million-Year Hike
PERE MARQUETTE STATE PARK
Jersey County

Nor has all this great work been without design; but, like all other works of the creator, we can see definite objects accomplished by it. And when we see a railroad following a valley of erosion . . . do we not see one of those objects?
—Edward Hitchcock, *Report on the Geology of Vermont,* 1861

The great Yankee geologist Edward Hitchcock shared the view of many of his contemporaries that the earth's landforms, like its plants and animals, had been created expressly for humankind's convenience—Vermont's valleys, for example, had been carved to eventually ease the task of building the state's first train lines. Using the same logic, I would like to suggest that through many ages the Good Lord molded Pere Marquette State Park for the latter-day delight of geologists and naturalists. At this one site, Illinois's largest state park, there is so much earth history, and so many rock formations and structural features, that several visits barely suffice. A single twenty-minute hike here leads you back through 120 million years of the long-vanished Paleozoic era.

The Goat Cliff Trail is well delineated and easy to follow, but because it crosses springy ground, ooze shoes are a must. You will quickly come to the first outcrop on the right, toward the bluff face. It is a large mass of gray, layered rock projecting above the slope. In most places in this state, the strata are flat lying or almost so. Are these beds more or less horizontal, in good Illinois fashion, or are they tilting? It doesn't take a geologist with twenty years' field experience to tell that the beds are dipping markedly toward the south. The formation here is the middle Mississippian St. Louis limestone, described more fully in essay 20. Take a close look at the lower part. It is a zone of breccia (pronounced BRETCH-*uh*). Breccia comprises angular fragments of rock, in this case the host limestone, that have been cemented together. According to a recent theory, the fragmentation was caused when the St. Louis was exposed to the open air in late Mississippian time and was afterward subjected to chemical expansion.

As you ponder what caused these beds to tilt, continue down the path. Before long you wind around the Twin Springs site, a second outcrop with

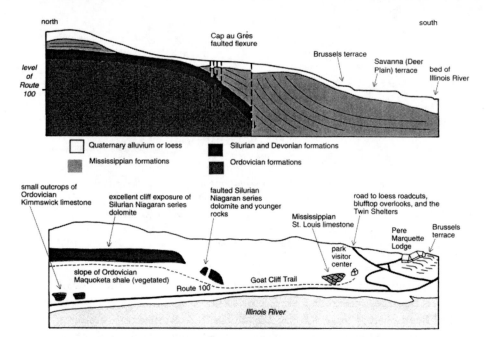

Two views of Pere Marquette State Park: (top) cross section showing the surface and subsurface structures and (bottom) a perspective view with outcrops and other features discussed in the text. The suggested hike, the Goat Cliff Trail, begins at the visitor center. —Illinois State Geological Survey

a distinctive solution-pitted surface. This is a marine carbonate rock, too, but it is a much older type: Niagaran series dolomite from an earlier Paleozoic period, the Silurian. This same suite of Silurian formations can be found in the Chicago area, and in part on the highest knobs of Jo Daviess County. In walking a short distance from the St. Louis limestone outcrop to this one, you have slipped backward about 60 million years. Since you did not go appreciably downhill to go downsection, the strata are obviously in an unusual orientation. Farther up on this assemblage, the Silurian beds are overlain by younger ones from the Devonian and Mississippian. All these beds, like those at the first outcrop, dip southward at a little more than 25 degrees from the horizontal. Furthermore, if you look closely you will see faults—giant stress cracks—that are situated almost perpendicular to the strata and displace them somewhat.

Geologists who have mapped this and other outcrops nearby have determined that this is one of the hinge points, so to speak, of a large structure known as the Cap au Grès faulted flexure. The faulting shows the strata in this location were pushed beyond their limit to bend without outright breakage. The flexure is the southern side of a huge arch-shaped wrinkle in the crust, the Lincoln anticline. Here at the Cap au Grès flexure, gently southward-dipping strata suddenly plunge more steeply, and in some cases straight

The Mississippian St. Louis limestone, along the Goat Cliff Trail just north of the museum, dips southward at more than 20 degrees.

downward. Because you have moved south to north crossing a part of this zone of dipping beds, you have been moving toward ever-older formations.

To find the oldest strata exposed in the park, leave the Goat Cliff Trail and carefully continue up the eastern side of Route 100 until you reach a small group of carbonate-rock outcrops close to road level. These are part of the Middle Ordovician Kimmswick limestone, a rock layer of economic importance in southern Illinois because it is a pay zone, or major oil-producing formation. It is the middle section of the Galena group (see essays 1 and 2). Above the Kimmswick at this site lies the Maquoketa shale, which, true to its nonresistant constituent rock types, forms a gentler slope obscured by colluvium and vegetation. You may find pieces of Maquoketa talus that have fallen close by.

Any ramble that covers this much of the geologic time scale is bound to be enlightening. But the main question still remains. What caused the Lincoln anticline and the Cap au Grès flexure to form in the first place? Apparently, they were the result of crustal stresses of a truly continental scale. The main episode of deformation has been tentatively assigned to the late Mississippian period—a time when the eastern edge of North America was well on its way to colliding with Europe and Africa to form the ancestral Alleghenian and Ouachita Mountains. Does this continental collision have anything to do with the folding here? Many geologists now believe so. This major structural feature, in common with the La Salle anticlinorium

The Pere Marquette State Park lodge sits on the edge of the Pleistocene Brussels terrace. To the right of the building, the surface slopes down to the younger Savanna terrace.

and Illinois's other zones of folding and faulting, may very well be a Midwestern response to the powerful compressive forces at work a thousand miles to the east in late Paleozoic time.

As suggested at the outset of this essay, there are many geologic high points in Pere Marquette State Park. If you resume your original track up the relentless slope of Goat Cliff Trail, you will pass an excellent cliff-forming outcrop of Silurian dolomite. Here, on the north side of the flexure, the Silurian rocks are exposed at a higher altitude than the ones you saw earlier at the Twin Springs site. While not exactly horizontal, these beds dip southward at a much shallower angle. Before long, you come to a scenic overlook facing north. Better yet, the trail angles back to a still higher vantage point, one of the park's hill-prairie shelters situated atop the bluff. When not distracted by the antics of cavorting bald eagles, you should be able to see another interesting geological feature across the Illinois River. The narrow finger of land between the Illinois and the Mississippi, Calhoun County, is part of the Lincoln Hills section of the Ozark Plateaus Province. It is also one of three sections in the Prairie State that probably was never glaciated. As you trace the level of the hilltops, you'll notice that they are all at the same general height. This now-dissected upper surface is a major remnant of the old Tertiary-period Calhoun peneplain, a thoroughly eroded, low landscape that was subsequently uplifted and subjected to a fresh cycle of erosion.

Even the park's elegant lodge, a favorite dinner spot both for local folks and residents of St. Louis, has a superb geologic context. The handsome

Pleistocene loess along the park roadway leading up to the Twin Shelters. Loess typically forms near-vertical faces.

stone and timber building sits on the edge of the well-defined Brussels terrace, which has traditionally been considered a former Illinois River floodplain level during the Pleistocene. In strolling riverward down the slope from the lodge, you descend to the lower Savanna (or Deer Plain) terrace, which in this area serves as the surface for Route 100. Another token of the Pleistocene can be seen as you ascend the park's blufftop road. A little before you reach the parking area for the Twin Shelters, a large roadcut reveals a windblown loess deposit that dates from the final stage of the Pleistocene, the Wisconsinan. The loess, silt originally brought to this region by the glacial meltwaters that coursed down the Illinois Valley, now forms a deep blanket on the blufftop. It has even created a unique modern habitat, the hill prairie, which is home to many rare plant and animal species.

If you stop at the Twin Shelters turnoff and make the short climb up to the scenic overlook, you will face three connected bodies of water. In the foreground is the slackwater environment of Stump Lake. Beyond Stump Lake lies the relatively narrow main channel of the Illinois; beyond that, the still waters of Swan Lake. In the early part of the Pleistocene, this valley, which marks the boundary of Jersey and Calhoun Counties, was used not by the present stream but by the region's master river, the Mississippi. It was only about 20,000 years ago that the Father of the Waters was diverted farther westward by the greatest extension of the Wisconsinan ice sheet. The smaller Illinois then appropriated this course. But here, this close to its junction with the greater and older river, who could say that the busy, barge-laden Illinois is not one of the country's great waterways?

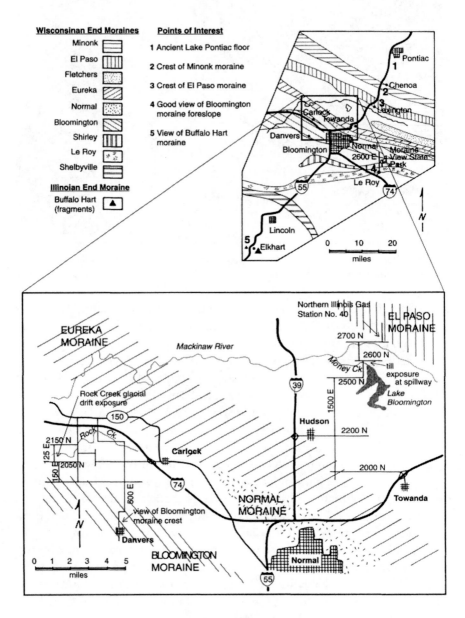

The Bloomington Ridged Plain. The tour begins at Pontiac and heads south on Interstate 55.

— 25 —
Moraine-Surfing Down the Interstate
PONTIAC TO LE ROY
Livingston and McLean Counties

Here the water went down, the icebergs slid with gravel,
the gaps and the valleys hissed, and the black loam came, and
the yellow sandy loam.
                                        —Carl Sandburg, "Prairie"

Each day thousands of motorists speed southward from Chicago on Interstate 55, trying their best to stay awake in the open stretches of downstate Illinois after successfully negotiating the traffic jams, gapers' blocks, and construction slowdowns of the Windy City's complicated highway system. The 250-odd miles between Joliet and East St. Louis can be a dreary ride for the geologically uninitiated; but to the person with an interest in the state's Ice Age legacy, the interstate is nothing less than a well-illustrated picture book of Pleistocene landforms. In the northern segment of the trip, knowingly or not, travelers participate in what I describe as moraine-surfing, when the superhighway leads them over a series of ridges that radiate outward from the Lake Michigan shore like expanding ripples in a pond.

This essay describes the physiographic province known as the Bloomington Ridged Plain. The name sounds like an oxymoron; but it is mainly flat ground, interrupted every so often by end moraines built by the Wisconsinan glacier near the end of the Pleistocene epoch, from about 20,000 to 15,000 years ago. There are more than a score of these low, arcuate ridges in this province. When traced on a map, their relation to the Lake Michigan basin, and to the lobe of ice that proceeded out of it, becomes obvious.

Our itinerary begins at Pontiac, a rest-stop town familiar to many interstate travelers. The route follows I-55 as far south as Bloomington, where it turns southeast on I-74 and proceeds to Le Roy. Before we scrutinize the individual points of interest, take a moment to ponder how end moraines came to be in this spacious prairie country. First, consider the mechanics of how the Pleistocene ice sheets, the source of all moraines, moved. The action began far north, not in the polar-sea ice pack, but on land, in the ice sheets' dome-shaped cores, or spreading centers. In eastern and central North America, separate spreading centers were located on both sides of

183

Hudson's Bay. There, large amounts of fresh snow accumulated until they were compressed into ice. The ice then expanded laterally outward. Where the ice traveled over frozen ground, it moved slowly ahead, in one bulge after another. In warmer locations, where the soil was not frozen, there was some bulging, but the glaciers were helped along by a low-friction layer of melted water or waterlogged debris at their base. The most important thing to remember about this whole process is that the glaciers had no reverse gear. They could not retract like a recoiling amoeba. Even when they were stagnant or receding they were still moving forward, in the sense that they were still being fed with fresh ice from the rear. But the increased rate of melting at the front—due to a net increase in summertime temperatures— prevented them from gaining any additional ground.

The leading edge of an ice sheet was the end of the line, so to speak, for the rock debris that had been locked in its base or carried along under it. For a sizable end moraine to develop, a complete standoff between advance and melting had to be sustained for a long time. Had it been otherwise, the rubble would not have been piled up along a ridge but would have been strewn over a much larger area of lower relief, to create a ground moraine instead of an end moraine. Ground moraines are good indicators that there were rapidly warming temperatures and a speedy meltback. It may have taken about one hundred years for a typical Wisconsinan end moraine to form, during which the ice would have remained stationary, delivering its load in conveyor-belt fashion. This estimate seems to fit in with the rest of the Pleistocene timetable, given the total number of Wisconsinan moraines and the other factors involved.

As a general rule, the farther one goes from the Lake Michigan basin, the older the end moraine is. Accordingly, the Bloomington moraine, which in this area is more than 100 miles from Chicago, was formed before the Tinley moraine, which lies less than 20 miles from the Loop. This makes sense when you think about it: if a new readvance of the ice extended farther than an older moraine, the older ridge would have been largely swept away. However, there are many examples of younger readvances that have partially overridden preexisting end moraines without obliterating them. In that case, the newer end moraine sits piggyback atop the backslope of the older structure.

Because the first leg of your exploration proceeds from northeast to southwest, you will be heading from younger to older moraines. In the vicinity of the Pontiac starting point, you will spot no ridges at all. In fact, this is one of the flattest surfaces anywhere in the state. What landforms are even more level than a Wisconsinan till plain? River floodplains are rela-

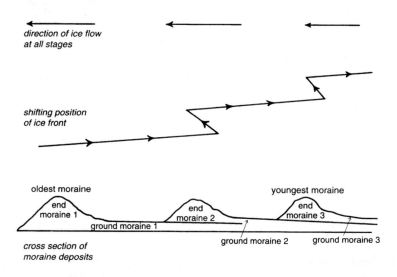

How things work at the margin of a moraine-building glacier. —Modified from Bretz, 1955

tively narrow features; the answer must be something like that but much broader. As remarkable as it may seem, you are driving on the bed of ancient Lake Pontiac. This large body of water was formed approximately 17,000 years ago, when glacial meltwater rose behind the natural dam of the Minonk moraine, to the southwest. Recent investigations have determined that the lake drained before the onset of the Kankakee Torrent, 15,500 years ago. While it lasted, however, Lake Pontiac provided a slackwater environment where silts and clays settled to the bottom. These fine sediments contributed to the development of an especially fertile soil. If you have any doubt that the Pleistocene and the Holocene have been times of rapid transformation, consider how the Pontiac area has changed in seventeen short millennia: from a basin swollen with turbid meltwater, to the ocean of grasses and wildflowers the American Indians knew, to the relentlessly rectangular crop-producing machine it is today.

The first real ridge you will see, a little more than 10 miles south of Pontiac, just after the Chenoa exit, is the Minonk moraine, a subdued feature approximately 3 miles wide at this point. Keep in mind that, like its companions farther down the road, it is not the equivalent of a line of snowclad mountain peaks. This trip is an exercise in subtle changes of relief. One reason the Wisconsinan ridges are low and wide is that they contain a large percentage of clay and silt the ice sheet dredged up from the Devonian New Albany shale, which forms the bed of Lake Michigan. Sometimes, however, the moraine crests are highlighted by tree growth and farm buildings that accentuate their relief. In the case of this low moraine, do not expect to see much of a well-defined ridge looming in front of you as you approach it. You'll know you're on it when you notice the characteristic swell-and-swale terrain. This easily recognized trait is an excellent way to determine you're on top of an end moraine, even if its features are otherwise subdued.

The next moraine, the El Paso, comes quickly. It is much more obvious, in large part because the town of Lexington occupies a strategic position along its crest. Its ridge rises into view about 2 miles after you descend the foreslope of the Minonk. The leading edge of the El Paso moraine fronts the Mackinaw River. The stream's southern bluff might be mistaken for another moraine, but in fact it rises to a relatively level till-plain surface. The next ridge, Fletchers moraine, lies on the eastern side of the road, 1 mile north of the Towanda turnoff and just before the highway crosses Money Creek. If you're in the mood for an enlightening countryside detour, get off at the Towanda exit and head north on 1900 East and connecting roads to the man-made Lake Bloomington. This area contains several excellent sites that the Illinois State Geological Survey uses as stops on its public field trips.

The most famous of these locales—famous, that is, to students of the Ice Age—is at the lake's northern tip, on the western side of the spillway. At this spot, glacial deposits from both the Wisconsinan stage and the preceding Illinoian stage were once extensively exposed. These days most of this cut is covered with vegetation, but if you carefully poke about (do not lose your footing) you can still get a feeling for what is underneath. The deposits at this site are primarily till, an unsorted jumble of sediments ranging from clay and silt to cobbles and boulders.

Just a mile to the northeast, on the other side of the Mackinaw River and therefore on the foreslope of the El Paso moraine, the Northern Illinois Gas Company has built Station No. 40—an underground natural-gas storage (but not production) facility. As you approach the site, you can usually smell its prime commodity. The gas is pumped more than half a mile straight down, into the Cambrian-period Mt. Simon sandstone, the rock section that sits directly atop the Precambrian basement. Because the Mt. Simon is capped by impervious Cambrian Eau Claire shale, and because the two formations are arched upward into an anticline here, the gas remains confined in one vast but invisible reservoir.

If you're feeling inquisitive and adventurous, you can also head over to the Carlock-Danvers area, northwest of Bloomington. The town of Danvers occupies a superb position atop the distinct Bloomington moraine, which stands at about 880 feet above sea level. Somewhat more difficult

This large streamcut along Rock Creek west of Carlock contains both Wisconsinan and Illinoian glacial drift, as well as an ancient soil zone from the Sangamonian interglacial.

to find is my favorite glacial-drift exposure, on the cut face of Rock Creek, 6 miles west of Carlock. The cut can be seen on the eastern side of 125 East, between 2150 and 2050 North; it is on private property, so do not trespass. From the road you can clearly discern the Wisconsinan Delavan till (the upper portion) and the pinkish Wisconsinan loess known as the Roxana silt (in the middle). Farther down, near creek level, the Illinoian Radnor till forms the base of the cut. Specialists who have studied this site have noted that the Radnor member has a relict soil profile, indicating it was at the surface during the warm interglacial stage between the Illinoian and Wisconsinan glaciations. This ancient, organically altered zone is known as the Sangamon soil.

When you return to Interstate 55 in the Bloomington-Normal area, you will be on terrain defined by two end moraines named after the twin cities. On the south side of the urban area, take Interstate 74 southeastward to the farm town of Le Roy. Exit there and head north on 2600 East, toward Moraine View State Park. The road brings you to one of the finest end-moraine vistas in central Illinois. Long before you reach the park access road entrance, you'll behold the long, high rampart formed by the Bloomington moraine. In this setting, it is easy to visualize the hulking mass of the Wisconsinan ice sheet standing along this line a little more than 18,000 years ago.

For many decades, geologists believed that the Pleistocene glaciers were, in some places at least, well over a mile thick. Recent studies of glacial-drift compaction, however, together with new research on the structure and dynamics of modern glaciers, indicate that the ice was probably 2,000 feet thick in the Lake Michigan trough, and only about 700 feet in more southern locations such as this. Even at 700 feet, it would have been as high as a 60-story skyscraper and it must have been a breathtaking sight.

If you would like to see an excellent and rare example of a much older Illinoian end moraine and are willing to go an additional one-way distance of about 60 miles, return to Bloomington and get back on Interstate 55 heading south. In Logan County, about 10 miles southwest of Lincoln, high ridgelike or moundlike segments of the Buffalo Hart moraine stand along the interstate at the town of Elkhart. In the relatively long span of time—125,000 years—since the end of the Illinoian stage, this feature has been breached and eroded in many places, but it can still be traced across four counties. The most impressive segment along the highway rises abruptly as an elongated hill. Its height puts many Wisconsinan moraines to shame.

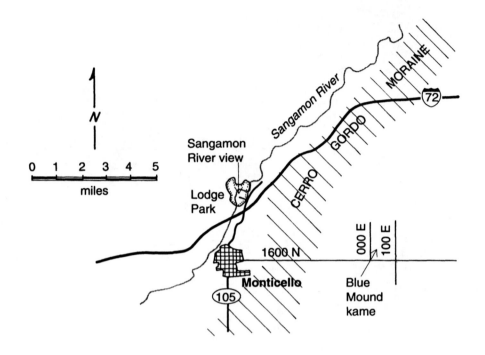

Site map of Monticello and vicinity.

A Tale of Two Rivers

MONTICELLO AND ENVIRONS

Piatt County

Altissima quaeque flumina minimo sono labi
(The deepest rivers flow with the least noise).
—Quintus Curtius, first century A.D.

Illinois is a land of great rivers, some of which are completely invisible. This odd fact is the result of the far-ranging, brutal work of the Pleistocene glaciers. Buried beneath the thick blanket of glacial debris lies a vast network of long-vanished streams. Few residents of the state know much about them, yet these unseen waterways have a tremendous impact on our everyday life, because they provide us with abundant supplies of that most precious natural resource—fresh water.

The greatest of all these now-buried streams was the Mahomet River. When I was a geology undergrad at a university in northeastern Indiana, I was taught that this immense ancient waterway was called what most out-of-state geologists call it: the Teays (pronounced *tays*). Within the Prairie State's borders, however, the alternative name Mahomet prevails, honoring the Champaign County town that stands above the river's deepest section.

If you could magically remove the burden of glacial drift from central Illinois, the Mahomet would stand out as this region's greatest landform, with a valley wider than the modern Mississippi's and a floodplain set 200 feet below the surrounding upland. And, if you could track the Mahomet to its source, you'd find that it extended upstream through northern Indiana, western and southern Ohio, West Virginia, and Virginia. Downstream from the Monticello area, it swung northwestward to meet the ancient Mississippi at the site of the Tazewell County town of Delavan. It would not be exaggerating to say that the ancient Mississippi was the Mahomet's tributary, not the other way around.

Here in east-central Illinois, the Mahomet valley bottom now sits on average more than 300 feet below the surface. Because much of the buried depression is filled with the Mahomet sand—a body of porous grits and gravels that were deposited in the early, Pre-Illinoian part of the Pleistocene epoch, the Mahomet makes a superb aquifer, or groundwater source. Ac-

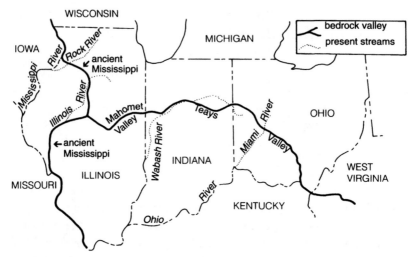

The ancient Mahomet-Teays River flowed through a great valley that traversed Ohio and Indiana and met the ancient Mississippi near the center of Illinois. —Illinois State Geological Survey

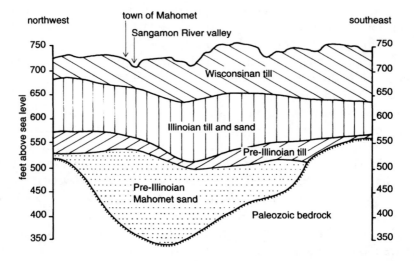

Cross section of the buried Mahomet-Teays River valley in the vicinity of Mahomet and Monticello. Note the size of the modern Sangamon River valley (cut into the Wisconsinan till) in relation to the size of the Mahomet-Teays River valley (cut into the underlying bedrock). The vertical scale is exaggerated for clarity. —Illinois State Geological Survey

cording to the State Geological Survey, in one year alone during the 1970s, the cities of Urbana and Champaign extracted five billion gallons of water from this seemingly inexhaustible source.

The Mahomet's bountiful water reserve does not wind its way underground from the western flank of the Appalachians to Illinois. (Nor does well water in northeastern Illinois come from Lake Superior, as many Chicagoland residents seem to think.) The sources of the Mahomet's water are local. Rainwater and snowmelt, percolating through the glacial drift, are picked up in the Mahomet sand, where it is retained until our ever-thirsty species pulls it out again.

As in every other reconstruction of the past, there is more than one plausible theory about the Mahomet's age and origin. Some geologists believe that the Mahomet-Teays system, including its Midwestern segment, was a Tertiary-period drainage net that predated the Pleistocene considerably. Others argue that the Tertiary Mahomet-Teays probably flowed northward to a Canadian outlet, instead of crossing the large Cincinnati arch bedrock structure to the east of Illinois. In this scenario, it took early Pleistocene glaciers to bend its course westward over the arch and through the Hoosier State. If this alternative is correct, the Mahomet entrenched itself in its central Illinois valley sometime after the first Pleistocene ice advances.

Whichever interpretation is true, here in Holocene time the grand Mahomet Valley is quite literally out of sight. On the north side of Monticello,

The Sangamon River winds through Lodge Park. In the Monticello area, this younger stretch of the river flows on top of the buried Mahomet Valley.

Many Wisconsinan end moraines are low features but are still recognizable as long ridges in the distance. The photo looks westward toward the crest of the Cerro Gordo moraine, which stands just east of Monticello.

though, in the lovely setting of the Piatt County forest preserve known as Lodge Park, you can see one of its successors. In the park's bottomland, somewhat obscured by swamp trees, flows the much more modest Sangamon River. It sets no records for immensity, even as it nears its junction with the Illinois River, far to the west in Beardstown. Nevertheless, there are two unusual aspects about this stream. For one thing, its valley here sits atop the Mahomet's—like a small child hoisted onto her mother's shoulder. For another, the Sangamon is a river with two different ages, with a headward section much younger than the downstream portion. The dividing line between these two sections lies a little west of Decatur.

The earliest phase of the Sangamon's history was the Illinoian stage of the Pleistocene (approximately 300,000 to 125,000 years ago), when meltwater from the ice sheet in central Illinois first formed the river. In the interglacial stage that followed—named, appropriately, the Sangamonian—the river entrenched itself in the state's midsection. When the Wisconsinan ice advanced into east-central Illinois, the upper segment of the Sangamon Valley was lost under a mantle of glacial drift. After the maximum Wisconsinan ice advance, the Shelbyville (about 20,000 years ago), the glacier resumed its retreat, with periodic small readvances. The meltwaters coursed down the surviving part of the Sangamon Valley, widening it considerably in the process. For that reason, the lower Sangamon now has a much broader lowland that the river's upper reaches.

As the Wisconsinan glaciation continued, ponded meltwater was held back by the euphoniously named Champaign moraine, northeast of Monticello. Finally, this proglacial lake rose high enough to cut through the

194

natural dam and carve the channel that became a large part of the new upper Sangamon. The Sangamon continued to work its way headward. Its main fork has even squeaked through the Bloomington moraine to reach its present source north of Moraine View State Park.

While you're in the Monticello area, take a quick look at two other local Pleistocene landforms. The first is the Cerro Gordo moraine, which marches down the eastern side of the Sangamon Valley from Mahomet to Monticello. You can get a good view of its foreslope as you approach Monticello from the east on 1600 North. If you take 1600 North in the other direction, you will see a low, gently sloping hill on the north side of the road, just over the Champaign County line between 000 East and 100 East. This hill, Blue Mound, is a kame, a type of ice-contact feature relatively common in Illinois. It formed when a stagnant glacier stood here. In one place in the ice, there was a depression or hole where sand and gravel accumulated. The ice melted away, but the sand and gravel remained to form this rise on the otherwise flat-as-a-chessboard till plain so typical of this region.

Because the till plain in this part of the state is very young (Wisconsinan), it is generally poorly drained—or at least it was until farmers put in ditches and drainage tiles. The tributaries of the Sangamon and other major streams have not had the requisite time to dissect the land the way those of the Illinoian Galesburg Plain, described in essay 19, have. Before the coming of big-time agriculture, the ground was often waterlogged, a condition favoring the growth of wet-prairie grasses. The growth and decay of generation after generation of those plants made this abundantly fertile soil, black as the night of a new moon, the envy of the world.

In the foreground, a classic example of Wisconsinan till plain, with no visible streams or gullies. In the distance rises Blue Mound, a kame. Photo taken east of Monticello.

195

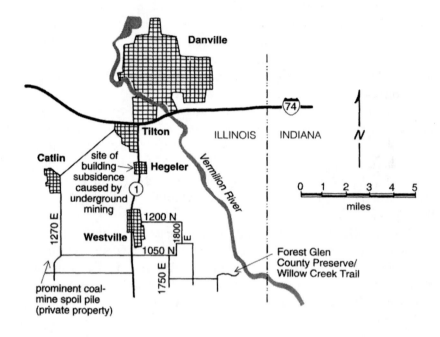

Site map for environs south of Danville, including Forest Glen County Preserve.

— 27 —
Coal Country and Pennsylvanian Cyclothems
FOREST GLEN COUNTY PRESERVE AND ENVIRONS
Vermilion County

> *The heterogeneous strata of the Pennsylvanian system in the central and eastern states constitute a complex succession so different from the older Paleozoic systems that the interpretation of the Pennsylvanian is difficult and few geologists have attempted more than a generalization of its geological history.*
> —Harold Rollin Wanless and James Marvin Weller, 1932

The first recorded discovery of coal deposits in North America was along the banks of the Illinois River, near the modern city of Ottawa, in the 1670s. The discoverers were two Frenchmen known to every Illinois schoolchild: Père Jacques Marquette and Louis Jolliet. Ever since the late seventeenth century, inquisitive human beings have been fascinated by the state's many coal seams, by the rock strata that surround them, and by the abundance of fossils they contain. Not surprisingly, the primary emphasis has always been on extracting the coal for profit. And great profits were there to be made. Coal fueled the Industrial Revolution and changed the civilizations of the world so quickly and so radically. Even now, two centuries after that revolution began, this black combustible rock remains one of Illinois's top natural resources.

Coal is by no means restricted to Pennsylvanian rock formations. In Australia, the mining of Permian-period coal is a major industry. That coal was formed from plant types different from the ones that grew here. Coal can come into being wherever vast amounts of plant debris accumulate in a swamp environment, form a peat layer, and suffer subsequent burial with sufficient heat and compression. In Illinois, and in the Midwest in general, it is the Pennsylvanian period that is the Coal Age. All the coal taken in this region is the bituminous variety, a softer substance than the more highly metamorphosed anthracite mined in eastern Pennsylvania.

The quest for expanded mining and profit has often wreaked havoc with the environment, but it has also spurred an increase in geologic research. Despite the intense interest in coal exploration and technological improvements, the significance of the Pennsylvanian strata taken as a whole eluded geologists for several generations. Wanless and Weller, quoted above, were Illinois-based researchers who did much to pioneer the way in deciphering the puzzling patterns of coal-bearing formations. Theirs was no simple task.

The Pennsylvanian period, 320 to 286 million years ago, was a truly strange episode in earth history, when conditions changed very quickly by geological standards. The record of the rocks shows that there were sudden shifts back and forth between marine and nonmarine conditions, and from thriving jungle habitats to those typical of sea bottoms far offshore. But it also seemed as though those changes were capricious, without a pattern that would point to a general principle. As Wanless and Weller noted, the sequences of rock types in any one county varied almost as much as they did across the whole state. But the two geologists succeeded in finding and describing a pattern that at least one earlier investigator, J. A. Udden, had alluded to. They did this by looking not just at individual localities but at the coal-bearing region as a whole. What they found in examining the larger picture was many variations on a single theme. Pennsylvanian rock types repeated themselves, the way a distracted person practicing the piano repeats a musical phrase over and over: often some of the notes—the individual rock types—are omitted, while others are inserted in unexpected places. It might be hard to recognize, but the basic musical phrase is still there.

Using that metaphor, the geological equivalent of the basic musical phrase is the cyclothem—a series of different rock types that succeed each other in a way that demonstrates a logical if rapidly changing set of conditions. The ideal Pennsylvanian cyclothem sequence begins at the bottom, with nonmarine rocks: a sandstone, followed by a sandy shale, a limestone deposited in fresh or brackish water, a gray layer known as the underclay, the coal itself, and a gray freshwater shale. Above that shale were the marine rocks: a limestone with fossils of saltwater organisms, then a black shale, then more limestone, and finally more gray shale. But to repeat, this is only the ideal; as suggested above, the individual layers are often missing or positioned out of sequence.

A particularly good place to see a portion of a cyclothem is south of Danville, in a large cut face along Willow Creek, in Forest Glen, the beautiful and well-maintained forest preserve administered by Vermilion County. Take the main preserve road past the headquarters building, downhill toward the Sycamore Hollow campus. Park in the small turnout in the low area just

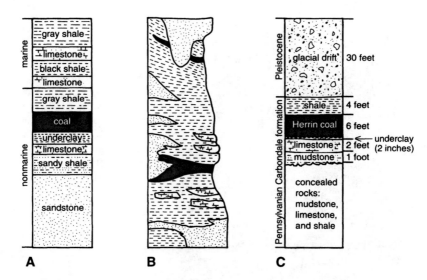

A: *A perfect Pennsylvanian cyclothem sequence.* **B:** *A real outcrop: some rocks occur as localized lenses, or interfinger with other layers; the ideal cyclothem sequence is interrupted or partially missing.* **C:** *The section for the Willow Creek exposure. The much younger Pleistocene drift is not part of the cyclothem sequence.* —Illinois State Geological Survey

below the main campus lot. The Willow Creek Trail starts across the road; you'll see the stream. As you set out on the trail, keep an eye on the creek's far bank. Before long you'll spot the Pennsylvanian Carbondale formation outcropping there. If you go down to the water's edge, you'll see that the rock has a network of joints, or vertical fissures, where no significant displacement has occurred. The joints here are also called syneresis cracks, because they probably formed as the waterlogged sediments dried out and shrank.

The real prize is located just around the big bend in the trail. Here a large streamcut reveals the following sequence: at the top, some Pleistocene drift; below it a shale member about 4 feet thick; a coal seam a little more than 6 feet thick; a very narrow zone of underclay less than 2 inches thick; 2 feet of reddish brown limestone with syneresis cracks; and a shalelike rock called mudstone. At times, this sequence is partially obscured by shale debris falling down from above. Nonetheless, with a little persistence you should be able to detect the large coal bed and the other rock types above and below it.

To geologists and miners, the large black swath exposed here is the Herrin coal (old-timers may still call it the No. 6 coal instead). The forest-preserve rules prohibit your collecting rock samples, but you can learn a

Willow Creek streamcut of the Pennsylvanian Carbondale formation at Forest Glen Preserve. Debris from the upper layer of crumbly, easily eroded shale has somewhat obscured the horizontal dark zone in the middle that marks part of the Herrin coal.

great deal by taking the time to carefully inspect the outcrop. It documents a few of the many changes the low-lying landscape experienced in the Pennsylvanian period. For example, the coal represents the tropical swamp forest and its lush growth of primitive trees and understory plants. (Keep in mind that in the Pennsylvanian, Illinois was just about on the equator, and North America was linked with Europe, Africa, and the other continents to form the great world-island of Pangaea.)

The shale lying over the Herrin coal signals the onset of the encroachment of water. In this case, according to the Illinois State Geological Survey's Colin Treworgy, the water was probably brackish, suggesting that a river valley was being drowned by the invading sea. If other strata has been atop the shale here, they might document the next step in the ideal cyclothem series: the presence of undiluted salt water. Again and again in Pennsylvanian time, land gave way to sea and sea gave way to land: the forests would rise anew in the steamy heat, only to be buried beneath the sea once again. The rapidly oscillating shoreline appears to have been caused by sea level changes produced by the destruction of ocean basins, by the rise and fall of ice sheets in colder regions, by fluctuations in the amount of sediments being delivered to the basins offshore—or by a combination of these factors. Wherever the coastline was at a particular time, the rivers continued to build deltas out into the sea as they shifted back and forth laterally across the coast.

In previous decades, the Danville area was one of the state's major coal-producing regions. The strip-mining technique was used for the first time a little north of this site, in Grape Creek, beginning in 1866. Underground mines were also prevalent. In a celebrated instance in 1967, the building housing a Hegeler radio station, together with some surrounding houses, sank about 3 feet into the ground as the mined-out area beneath them partially gave way. Large mining companies were not the only excavators of coal. If you continue up the Willow Creek Trail a little farther, you'll see a small entryway into the hillside. This is the adit of a small, one-person coal mine (in local parlance, a dog mine). In this coal-rich locale, farm families often dug dog mines to supplement their fuel supplies and income. Interestingly, the design of this tiny operation is a good example in miniature of the larger commercial drift mines that were tunneled into a hill or bluff from the place the coal was visible aboveground.

On your departure from Forest Glen, you may wish to explore the mined-out areas around Danville. The spoil piles of abandoned underground mines south of Catlin are famous for the excellent fossils found in their ironstone concretions. But as the many NO TRESPASSING signs indicate, you are not allowed onto the piles or the ground surrounding them without the landowners' permission.

The adit of a one-person dog mine. This site is just up the trail from the large Willow Creek cyclothem exposure.

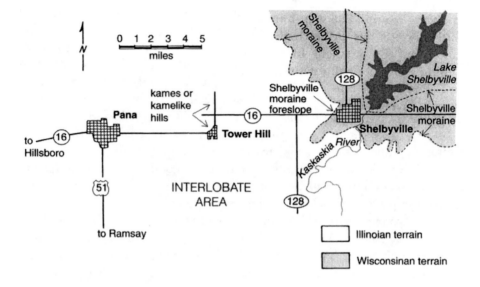

*The Shelbyville moraine and environs. The tour begins on Illinois 128
heading south toward Shelbyville.*

On the Edge of Two Ages
SHELBYVILLE AND THE INTERLOBATE COMPLEX
Shelby County

*We are now going to pass to another group of facts which prove
that the ice had, at an earlier epoch, a still greater extension.*
—Louis Agassiz, *Études sur les glaciers,* 1840

A century ago, when the glacial landforms of Illinois were first being
mapped, it was the custom to name Pleistocene end moraines after towns
where they were clearly visible. The Bloomington moraine passes obviously
through Bloomington; the Marengo passes visibly through Marengo; and
so on. In the same way, the town of Shelbyville has won a measure of notoriety,
in earth-science circles at least, merely by sitting atop one of the most
significant end moraines of all.

What makes the Shelbyville moraine special is that it marks the maximum ice advance in the final, Wisconsinan stage of the Pleistocene. That
advance stretched two-thirds of the way from Chicago to St. Louis, and
even managed to dislodge the ancient Mississippi River from its initial middle-
of-the-state course to its present valley much farther west. Even this greatest
extension of the Wisconsinan glaciation was nothing compared with the
advances of the preceding Illinoian stage, when the ice sheet swept down
to within 25 miles of Illinois's southern tip. It is the Illinoian ice that deserves
much of the credit, or the blame, for the fact that 85 percent of the state
was glaciated.

Because the Shelbyville moraine is the outermost rampart of
Wisconsinan landforms, the area around its namesake town is an excellent
locale to see the differences between Wisconsinan and Illinoian terrain.
Begin by approaching Shelbyville's outskirts from the north, on Route
128 south of Dalton City. Here the upper Kaskaskia River has been dammed
to produce one of the state's big artificial lakes. The upland surrounding
it is flat and devoid of streams—it is quintessential Wisconsinan till plain.
Because the ice sheet left this area late in the Pleistocene, the tributaries
of the Kaskaskia have not worked their way headward to the point that
the surface is highly dissected.

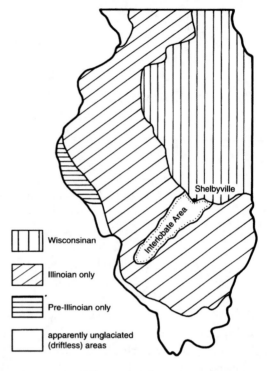

The Interlobate Area and its relation to the rest of the Illinoian glacial terrain.

Shelbyville

Interlobate Area

| | Wisconsinan

| | Illinoian only

| | Pre-Illinoian only

| | apparently unglaciated (driftless) areas

About 5 miles before you reach Shelbyville, you'll see that the road rises gently; you are now gradually gliding up the end moraine's backslope. In this area, the moraine has a maximum relief of only about 50 feet. Closer to our state's eastern border, it can be three times that high. The somewhat steeper foreslope, now mantled in shopping centers, fast-food restaurants, and other formations of the late Consumer period, comes to an end before you're out in the country again. But the farther you head out of town, the more obvious it is that the landscape has taken on a different aspect. A few moments ago, you were on a surface that is a mere 20,000 years old; now you are driving across one that is probably quite close to ten times that old. In physiographic terms, you've left the Wisconsinan Bloomington Ridged Plain and are now on the Springfield Plain of Illinoian age. The Springfield Plain is characterized by—some would say, notorious for—its relentless low relief. In that sense, it is reminiscent of the younger Wisconsinan till plain, but it is a much better drained province of numerous, relatively shallow streams that have not produced ravines as deep and sharply defined as those found in the Illinoian Galesburg Plain to the northwest.

On reaching Shelbyville's main intersection, head west on Route 16 toward Pana. When you make the southward jog into Tower Hill, the land-scape does the unexpected: the Springfield Plain suddenly gives way to an

Approaching the town of Shelbyville from the north. Route 128 here rises very gradually up the backslope of the Shelbyville moraine. When the moraine crest is this subtle, it's best to look for the swell-and-swale topography that is a better indicator of morainal terrain.

This kamelike rise, Tower Hill in the town of Tower Hill, is one of many such features in and about the Interlobate complex.

undulating, hilly terrain. You are entering what I call the Great Zone of Overt Lumpiness, better known to the geological community as the Interlobate complex. Beginning here, close to the outermost tongue of the Shelbyville moraine and extending roughly 80 miles to the southwest, is an elongated stretch of scenic high ground 20 to 30 miles wide.

If you take a short diversion on the side roads north of Tower Hill, you will spot fairly conical features that sit on north-south ridges or on the till-plain floor nearby. To the student of glacial geomorphology, a conical land-form shape is at once suggestive of the stagnant-ice features known as kames. Some of these hillocks are made of stratified sands and gravels, just as kames should be. However, some features are made of something else, glacial till. Usually, till is the hallmark not of dead ice but of a glacier that is still actively bringing fresh rock debris up to its front edge.

When you reach Pana, you have a choice: you can swing due south toward Ramsay on Route 51, or you can continue on Route 16 southwest-ward to the beautifully situated town of Hillsboro. Either way, you will be in a premium landform-spotting precinct of the Interlobate complex. You will easily discern the long ridges and, here and there, more kamelike struc-tures as well. As suggested above, this is a geologically puzzling area. Some investigators have theorized that the ridges are long Illinoian moraines; others propose that they are large, stagnant-glacier crevasse fillings. The second of these theories has a lot going for it, given the ridges' similarity with proven crevasse features, and given their preponderance of sand and gravel. What is so perplexing, though, is that the ridges are also mantled with deposits of till. If one switches back to the first theory, fresh problems arise. If these ridges are end moraines, why are they so straight and parallel to one an-other? Moraines typically have a crescentic or bowed shape.

The clues may be mystifying in this matter, but they are clear-cut in another. The evidence indicates that the ridged area is a sort of no-man's-land between the two great sections of Illinoian ice that invaded the Prairie State. The northwestern section appears to have been the more hyperki-netic of the two. Apparently, it advanced and retreated more than the south-eastern one. But of the Pleistocene-epoch ice advances in North America, Europe, and Asia, it was the southeastern section, in its maximum extension here in this state, that reached the lowest latitude. As it approached the northern flank of the Shawnee Hills, it established its unassailable claim as the champion long-distance runner of the Ice Age.

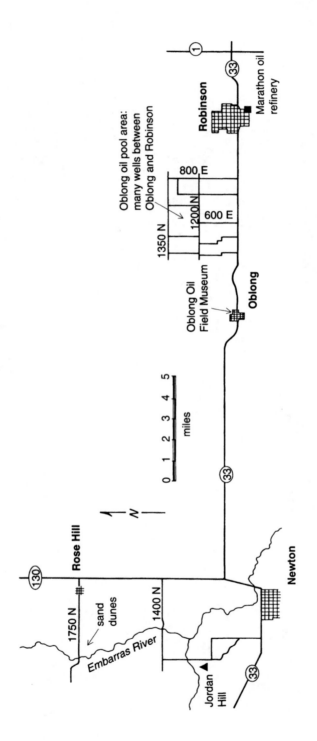

Site map of the oil country from Robinson to Rose Hill.

— 29 —
The Black Bounty of the Illinois Basin
ROBINSON TO ROSE HILL
Crawford and Jasper Counties

Present-day accumulations of petroleum and natural gas are found invariably in or adjacent to sedimentary rocks. According to all available evidence, these fluids have originated from the organic matter deposited in sediments—principally marine—at the time of their deposition.
— Marion King Hubbert, 1953

Few geologists today would argue with this explanation of the origin of petroleum offered by the University of Chicago–trained geologist who was one of America's foremost experts on oil exploration. It took Immanuel Velikovsky, the scientific world's favorite enfant terrible, to suggest that in the beginning our planet's stock of petrochemicals rained down from outer space. This picturesque alternative theory of the earth as a cosmic sponge lacks one key ingredient: supporting evidence. Still, when most people think of crude oil, they do not think of its geological underpinnings but of simpler images—the Persian Gulf emirates, the expansive oil fields of Texas, or the controversial Alaskan pipeline. Most people would regard Illinois as not exotic enough for oil wells. But for more than a century it has been a significant production center for both oil and natural gas.

In the United States, petroleum production began in 1859, when Edwin Drake drilled his famous well near Titusville, in western Pennsylvania. It was not long until the rush to find and market petroleum reached the Prairie State. This hunt was often carried out by skin-of-the-teeth, wildcat outfits that relied more on the luck of the draw than on sound analyses of the underlying bedrock. (My favorite of these prospecting concerns is the Impromptu Exploration Company, a name of refreshing candor.) Through all the bad choices and misunderstood geology, human beings had a way of finding what they were looking for.

To the nongeologist, a map of the state's many oil deposits looks something like jots of ink spattered across a blank page. Most of the sites, though, are clustered inside a giant structure known as the Illinois Basin. This vast accumulation of Paleozoic-era sedimentary strata, some 3 miles thick at its deepest point in the state's southeastern quarter, is shaped like the bowl of

The Marathon oil refinery in Robinson.

a spoon and is oriented along a northwest-to-southeast axis. Layer after layer of marine sediments, together representing a span of as much as 300 million years, settled on the gradually subsiding seafloor. The long record of deposition stretched from the Cambrian period to the Pennsylvanian or even the Permian period. Then, in the succeeding Mesozoic era, the rise of the Pascola arch below the state's southern tip separated the basin from surrounding structures and helped define its shape. The rocks get older deeper under the basin's surface, until at last the lowermost Cambrian strata give way to the igneous rocks of the Precambrian basement. On the surface, the older formations lie on the margin; the youngest surviving ones, which are Pennsylvanian in age, are in the center. The Permian beds, if they ever existed, have been entirely removed by erosion.

Oil wells are one of the most common sights you'll encounter on a meandering drive through the territory of the Illinois Basin. Oil-field pumps are particularly prevalent along Interstate 64 from Mount Vernon to the Indiana border. But an even better area to see active oil fields is farther north, on the flat till plain between the towns of Robinson and Oblong. Begin your westward trek in Robinson, and by all means take a good look at the imposing Marathon Oil Company refinery on the town's south side.

210

A common sight in the oil fields of the Illinois Basin: a well and pump (left) and oil storage tanks (right).

Here raw petroleum, a bituminous black liquid derived from the break-down of marine organisms, is subjected to a complicated distillation process that creates the lifeblood of modern civilization. The hustle and bustle around this facility, and the large number of active wells in the region, belie the fact that the domestic oil-production industry is still reeling from the low price of Middle Eastern petroleum.

Head west on Route 33 from Robinson toward Oblong. Before you reach the town of Oblong, take a quick detour north on the unpaved county roads to see the extent of the Oblong oil pool, which is one portion of the Main Consolidated Oil Field. The Main Consolidated stretches across 65,500 acres, including much of Crawford County and parts of Lawrence and Jasper Counties. Petroleum was first discovered here in 1906, and since then more than 245 million barrels of oil have been produced, from fifteen different pay zones (drillers' lingo for petroleum-bearing rock) in strata ranging in age from Ordovician to Pennsylvanian. The depth of the pay zones varies from a mere 510 feet to 3,900 feet. Here you will get a close look at the most common individual structures of a producing well: the rotary pump, bobbing up and down, and the upright cylindrical storage tanks in which the petroleum is stored until it is collected by a refinery hauler. You may

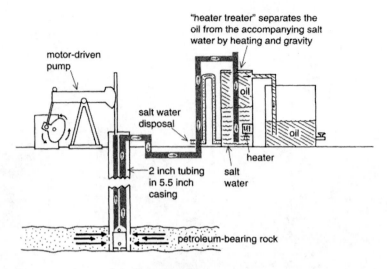

A schematic drawing of an Illinois Basin oil pump and its associated equipment. —Illinois State Geological Survey

also spot other well components, including the "heater treater" structures that separate petroleum from the salt water that accompanies it, and the open-flame flares that burn off natural gas from wells devoted solely to petroleum production.

As you drive about this flat expanse, try to imagine the subsurface conditions that provide the proper environment for petroleum accumulation. Anyone who has seen an oil slick in a harbor or a dab of butter floating in a water-filled saucepan understands that oil is less dense than water; it rises to the surface whenever possible. It is only when the strata form a trap—as when they are arched into an anticline, with an impermeable section capping a porous one—that oil is forced to accumulate into a pool underground. In the Oblong pool, the pay zone comprises three sandstones of the lowermost Pennsylvanian. Collectively, they are known by the local, informal name of the Robinson sand. As productive as these strata seem to be here, Pennsylvanian rock zones are by no means the most important source of oil statewide. Many wells have been sunk into Ordovician, Silurian, Devonian, and Mississippian beds. Wells generally succeed in extracting only a third or less of the petroleum in any particular rock layer. Consequently, petroleum engineers have developed the simple yet ingenious technique of waterflooding—pumping water into the pay zone. The water forces out more of the residual oil, but the maximum yield is usually only about one-half of all the petroleum present.

212

On your return to Route 33, continue west into Oblong and visit the Oil Field Museum, located just below Park Lake, on the northeastern side of town. This small but informative facility has a collection of old-time drilling and well gear. (To make sure it will be open when you get there, you can call the Oblong mayor's office before you set out.) You can range farther afield, both in geography and subject matter, by exploring the wide valley of this area's main river, the Embarras. (If you're not from around here, beware. The stream's name is pronounced *AM-braw*.) Its spacious valley is floored with Pleistocene glacial outwash and the more recent Cahokia alluvium.

In Wade Township, about 4 miles northwest of Newton, there is an interesting tree-covered landform named Jordan Hill. Visible west of 900 East just above 1225 North, it is the largest detached section of what was once a continuous ridge. Late in the Pleistocene, the ridge was subjected to the erosive force of the Embarras on one side and the local tributary, Turkey Creek, on the other. Eventually, the streams succeeded in punching a hole through the ridge; now Jordan Hill pokes up from the flat bottom-land like the top of a giant's head pushing out of the earth. In modern times, the Embarras flows a full mile to the east, where it has deeply incised itself into the unconsolidated valley fill.

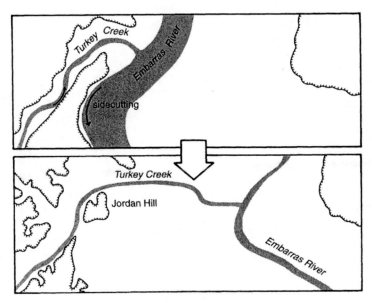

Sidecutting by the Embarras River created Jordan Hill late in Pleistocene time. —Illinois State Geological Survey

213

The forested mass of Jordan Hill, viewed from 900 East. Before it was divided by sidecutting erosion from the Embarras River and Turkey Creek, the hill formed a continuous ridge that extended across the view to the far left. Note the extremely flat river floodplain in the foreground.

Work your way farther northward, either on Route 130 or on the county roads paralleling it, to the small settlement of Rose Hill. There head west on 1750 North, back toward the Embarras's bottomland. Just before you descend into it, you will come upon a nice collection of dunes that were formed of outwash sands blown to the blufftop by the predominantly westerly winds. Many of these dunes are the crescent-shaped form known as barchans. Here more than one family has used the well-drained substrate and the high ground to site their homes. If you look closely, you'll see that the windborne sediments are not completely uniform. The sand that fell closest to the river is relatively large-grained, but a little farther east it is finer, and eventually grades into loess.

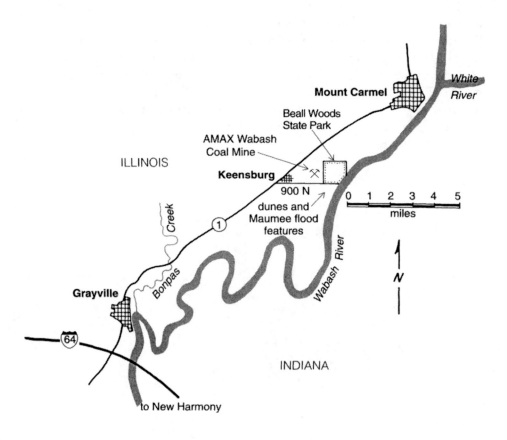

Beall Woods State Park is east of Keensburg on 900 North.

30

The Lower Wabash Considered
BEALL WOODS STATE PARK AND ENVIRONS
Wabash County

We have a multitude of instances of . . . the filling or increasing of the Plains or lower Grounds, of Rivers continually carrying along with them great quantities of Sand, Mud, or other Substances from higher to lower places.

—Robert Hooke (1635–1703)

As seen from high orbit, North America looks nothing like our political maps. The landscape manifests many variations, but there are no evident human boundaries. The separate countries and regions melt into a seamless entity. This fact reminds us not to think too narrowly. In Illinois, we must remember that the forces of nature hardly begin or end at the state's borders. Ice sheets from Canada, rock debris from Wisconsin, and ancient seas encroaching from the Deep South have all been crucial factors in the state's geologic history. One preeminent example of this need for a regional approach is found in the Illinois portion of the lower Wabash Valley. The evolution of this area can be understood only if we look far to the northeast, to the upper portion of the Hoosier State and beyond.

Start your exploration of the beautiful lower Wabash River country at one of its most famous sites, Beall Woods State Park. (In case you're not aware of it already, the first word of the park's name is pronounced *bell*.) To botanists and birders, the site's holiest of holies is its 290-acre nature preserve. Some authorities say it contains the largest parcel of near-virgin, broadleaved woodland east of the Mississippi River. Even though this claim has been disputed, the preserve is remarkable both for the large number of native tree species present and for the great age and size of its individual trees. For the earth-science enthusiast, there are many other points of interest on the state-owned property and just outside it.

If you have never beheld a modern underground coal mine, you'll have an opportunity to redress the deficiency when you approach the state park on 900 North, from Keensburg and Route 1. The imposing complex on the north side of the road is the preparation plant and associated structures of the AMAX Coal Company's Wabash Mine. The facility, which provides the fuel

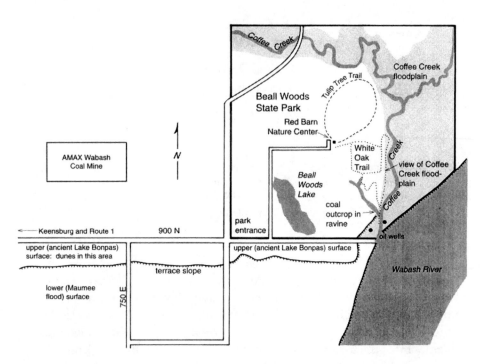

Beall Woods State Park and environs. The lightly shaded region represents the Coffee Creek floodplain.

for the power plant across the river in Gibson County, Indiana, began operation in the early 1970s and has been extracting the Harrisburg (or No. 5) coal. In this locale, the Harrisburg is more than 6 feet thick, and it lies approximately 800 feet beneath the earth's surface. In the future, the shallower Herrin (No. 6) coal, and another seam even deeper than the Harrisburg, may also be mined. (Perhaps you're wondering how, in light of the Clean Air Act, electric-power plants can still be burning coal. The answer lies not only in the economic incentive for doing so but also in the plants' smokestack-mounted scrubbers, which are designed to reduce sulfur-compound emissions.) What you are seeing here is like the unsubmerged part of an iceberg: the real business end is far out of view. The mine's underground working face has extended farther than you might think; for instance, it has reached under the irreplaceable natural treasure of Beall Woods. This is because the state was unable to obtain the subsurface rights to the park.

While you are still abreast the AMAX property, keep an eye out for the T-intersection of 900 North and 750 East. Turn south down the latter. Here, you are about half a mile from the bank of the modern Wabash

218

View near the entrance to Beall Woods State Park. The low area in the foreground is the Maumee Flood surface; the well-defined terrace above and behind it is part of the ancient floor of ancient Lake Bonpas. The AMAX Wabash Mine (right background) also occupies the high terrace.

River. You are also descending from an upper surface to a lower one, which was formed during a catastrophic event called the Maumee Flood. To put that dramatic point of prehistory in its proper context, we would have to turn the geologic clock back to late Pleistocene time, approximately 20,000 years ago, when the maximum extension of the Wisconsinan ice sheet created the long Shelbyville moraine (see essay 28). This ridge of glacial till crosses the state line far north of here, in the vicinity of Terre Haute; so it is clear that the Wisconsinan glacier did not reach the Beall Woods area, as the earlier Illinoian glacier most certainly had. But glaciers produce dramatic effects far beyond their direct reach. When the ice melted back from the Shelbyville front, it deposited a vast quantity of outwash—waterborne sand and gravel—as valley-train deposits along the lower Wabash. The amount of outwash was so great that the river became a braided stream, with many small channels winding through exposed gravel bars they could not remove. Part of this braided pattern is still visible in the valley-train deposits upstream near Vincennes.

The outwash also clogged the mouths of the lower Wabash's tributaries. Bonpas Creek, which now debouches into the big river at Grayville, backed up as far north as Richland County and formed ancient Lake Bonpas, one of several Pleistocene lakes in southern Illinois. Subsequently, however, much of the Wabash valley train was removed. About 14,000 years ago, the retreating Erie lobe of the Wisconsinan glacier formed a proglacial meltwater lake that stretched from the ice sheet's edge north of Ohio all the way over to northeastern Indiana. This body of water, Lake Maumee, continued to rise until the restraining morainal dam at Fort Wayne was overtopped and breached. In a way reminiscent of Illinois's even larger Kankakee Torrent, the lake's waters surged down the length of the Wabash Valley, scouring out about 20 feet of the valley train in the process. The difference in elevation between the upper and lower surfaces marks the depth to which the Maumee Flood cut through the outwash.

When you've taken a good look at the lower, Maumee Flood surface, retrace your way to the T-intersection. The upper surface along which 900 North runs is the exposed bed of ancient Lake Bonpas. In the vicinity of the intersection, it is dotted with hummocks of sand. Can you spot these dunes? They are formed of valley-train material that was exposed to wind action when the rushing waters of the Maumee Flood abated.

With that slight detour accomplished, continue eastward on 900 North past the Wabash Mine and enter Beall Woods State Park. Proceed to the end of the main roadway and park in front of the Red Barn Nature Center. Besides being an enlightening education stop, the building is the terminus for several top-notch nature trails through the first-growth woodland. Of these, I especially recommend the upland Tulip Tree Trail, because it features lofty specimens of *Quercus shumardii,* the Shumard oak. Similar to the bottomland-dwelling pin oak, the Shumard has a characteristic flared trunk base. Why should this concern a geology-centered visitor? Because the tree was named for Benjamin Shumard (1820–1869), a geologist and paleontologist who did fieldwork in the Midwest, Texas, and the Pacific Northwest. During the early part of his relatively short career, he was an assistant to David Dale Owen, son of the founder of Indiana's communal New Harmony settlement, located less than 20 miles south of here. Owen was one of the towering figures of early American earth science.

A stroll along White Oak Trail and down to Coffee Creek is a worthwhile exercise because it gives you an excellent opportunity to examine the channel and floodplain of this Wabash tributary. Look for the short feeder streams, which descend steeply and have deep gullies: this is a sign that they are still trying to adjust to the lowered base level of the modern Wabash. In times of high water, Coffee Creek swells rapidly and covers the surrounding swamp floor. If you happen to be here after such a flood, you will be able to find fresh sediments that have just been deposited on the floodplain. They are indicative of slack or slowly moving water, where the finer particles—silts and clays—can finally settle out of suspension.

Most of the time, Coffee Creek keeps within its banks and allows hikers to wend through an awe-inspiring bottomland grove that dwarfs the human form. At low water, you can even see Pennsylvanian strata exposed on the stream's banks. Before long, you'll come to the trail's bridge. Look carefully in the steep ravine nearby: the Pennsylvanian outcrop there contains a seam of coal. After that, you soon reach the Wabash. It's not the rip-roaring waterway it was in its Maumee Flood phase, but it deserves its reputation as one of America's major rivers. Not far from its near bank stand oil pumps—one, on the far side of Coffee Creek, is mounted on a platform to protect it from periodic floods. Here the underlying Pennsylvanian bedrock provided a modest pay zone.

The flared lower trunk of a majestic Shumard oak, a species named for a noted nineteenth-century American geologist. Photo taken along the Beall Woods Tulip Tree Trail.

221

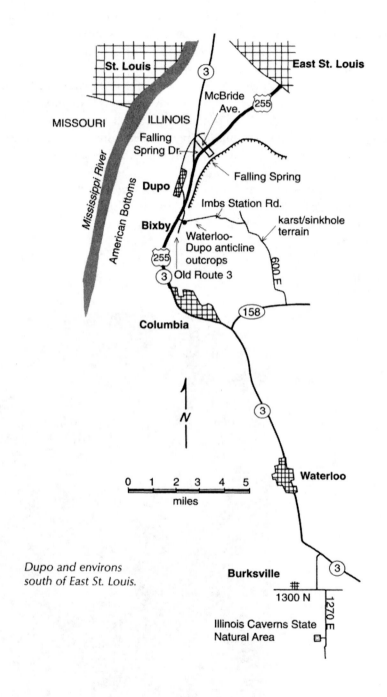

Dupo and environs south of East St. Louis.

31

Caverns, Sinkholes, and Falling Water: The World According to Karst

DUPO TO ILLINOIS CAVERNS STATE NATURAL AREA
St. Clair and Monroe Counties

I was as interested in the discovery of limestone as if it had been gold, and wondered that I had never thought of it before. Now all things seemed to radiate around limestone, and I saw how farmers lived near to, or far from, a locality of limestone. . . . I read a new page in a history of these parts in the old limestone quarries and kilns where the old settlers found the materials of their houses.

—Henry David Thoreau, journal entry from 1850

Probably no other rock type can match limestone's many uses. It provides lime for cement and other industrial applications, soil sweetener used in agriculture and gardening, aggregate for road building and other construction needs, and ornamental facing stone for buildings. It also serves as an ingredient in soaps, cosmetics, paint, and—believe it or not—bubble gum. Even when it is left to its own devices, limestone has a way of expressing its unique attributes. If conditions are right, it forms one of the strangest and most secretive of landscapes, characterized by sinkholes, subterranean streams, and bat-haunted caverns. Geologists call this kind of terrain karst topography. In Illinois, it can be found in several locales. The most extensive is in the Mississippi Valley fringe south of East St. Louis. Some experts might quibble about whether full-fledged karst has developed in the state. If we accept the plausible definition found in the current edition of the *Dictionary of Geologic Terms*, though, it truly has.

Karst comes into being where a region is underlain by a substantial quantity of carbonate rocks and where there is ample rainfall. The rainwater, rendered slightly acid by the carbon dioxide in the atmosphere, seeps into the ground and dissolves sections of the alkaline limestone or dolomite. The solution cavities form best when the rock is beset with a system of vertical joints and horizontal fissures between strata, and where it is fairly massive—that is, neither thinly bedded nor porous. These conditions promote the motion of water along the cracks and spaces, which is the starting point for the formation of tunnels and caverns.

The massive bluff face at Falling Spring. The top of the cliff is mantled in Pleistocene loess. Below it are the Mississippian Ste. Genevieve and St. Louis formations. The wavy disconformity in the St. Louis beds extends horizontally just below the middle of the photo, at the boundary between light and dark beds. Small clumps of vegetation help to highlight the disconformity.

The subject of karst origin kindled a lively debate in the geological community for decades. The two leading explanations were the so-called vadose and phreatic theories. The first postulated that the water creating the caves and underground streams does its work above the water table—above the uppermost level of a region's groundwater. Further, the theory said that the water is moving purely in response to gravity; eventually, it will emerge to meet a stream in some deep valley nearby. The phreatic explanation stated the opposite: the solutional activity takes place in the groundwater zone, where the dissolving water moves down or even up because it is under hydrostatic pressure. Nowadays, the debate is not considered particularly important, since most caves have gone through both phreatic and vadose phases of development. Karst caves these days are classified as epigenic (created near the earth's surface, as in Monroe County) or hypogenic (created at greater depth, as in New Mexico's Carlsbad Caverns).

In the Illinois portion of the greater St. Louis area, you can see a host of karst features, with a few other geologic marvels thrown in at no extra cost, in the space of a pleasant half-day drive. (Please note, however, that the final stop cited in this essay is an unlit cavern complex that you can visit only if you are equipped with spelunking gear, including a hard hat and portable illumination device.) Begin at Dupo, where once there were almost three hundred wells extracting petroleum from the Ordovician Kimmswick limestone (seen in outcrop in essay 24). Early in its development, this area was a prolific producer, but it was by no means one of the state's great oil fields. The community of Dupo sits on an especially broad section of the Mississippi River lowland known as the American Bottoms. The great waterway has sliced a wide swath—sometimes more than 10 miles wide—through the unresisting Pennsylvanian clastic rocks that once stood here. Just south of Dupo, the river meets Mississippian limestones that haven't given up the fight so easily. The valley there is more constricted.

The first point of interest is situated on the valley's eastern bluff line, at the foot of North Dupo's Falling Spring Drive. Here stands a beautiful high cliff exposing two Mississippian carbonate formations: the Ste. Genevieve, occupying the top 20 feet or so, and the more extensive St. Louis, below it. A small waterfall, issuing from a solution cavity in the St. Louis strata, tumbles onto a deposit of tufa, a black, spongy, almost mosslike substance formed when calcium carbonate in mineralized water precipitates out of solution. As you will see, this scenic spot is a favorite with the local population. It was also a preferred hangout for the native peoples of the Archaic period of pre-Columbian culture (8,000 to 500 B.C.).

The boundary between the two formations at Falling Spring is subtle, but if you remember that the St. Louis starts about 20 feet down from the clifftop, you should be able to trace the contact. Up at the top, look for striking crossbedding patterns in the Ste. Genevieve limestone. Such large-scale crossbedding is indicative of high-energy environments—exposed dunes, submerged surf zones, and so on. As a result, this type of crossbedding is more common in sandstones and other coarse-grained clastic rocks. The reason the Ste. Genevieve beds exhibit this high-energy hallmark is because they are made of a special granular limestone that apparently formed in the tidal zone close to shore. (This unusual rock type, oolite, is more thoroughly described in essay 16.)

Roughly halfway down the cliff, well within the St. Louis formation, you should be able to make out a zone of buff-colored dolomite, limestone's magnesium-rich equivalent. At the top of the layer from which the waterfall issues is a superb, grossly undulating disconformity, with up to 10 feet of relief. This irregular surface has been described as the result of highly erosive, nearshore wave action that worked on the upper section of the dolomite when it was the uppermost layer. Some Illinois geologists doubt this interpretation, however.

Next, make your way a few miles south, on Interstate 255 or Old Route 3, to the center of the small settlement of Bixby. At the intersection of Old Route 3 and Imbs Station Road, turn east on the latter. A creek parallels the road on your right. At precisely 0.25 mile up from the intersection, you will see a small outcrop of the St. Louis formation poking out of the creekbed. Here is a wonder to behold: the limestone strata, in gross disrespect for general Illinois custom, are not horizontal, but dip steeply to the west. Now head up the road a little farther. Beginning at 0.7 mile from the same intersection, Old Route 3 and Imbs Station Road, you will find more out-crops of the same formation. They too are dipping, but less steeply and in the opposite direction, to the east. Can you come up with a theory to explain this sudden change in the orientation of the strata? What you are seeing is the Waterloo–Dupo anticline, an archlike warp in the crust that runs north by northwest from the vicinity of Waterloo, Illinois, to St. Louis. Anticlines are a common occurrence in this state, but rarely do they express themselves so blatantly in public view. Here, along Imbs Station Road, you have traversed the arch in cross section. At the first outcrop, you were on its western flank, or limb; then you crossed the crest of the warp to reach the second set of outcrops, which are part of the anticline's eastern limb.

East of the crest of the Waterloo-Dupo anticline, the St. Louis beds dip gently eastward. Compare with photograph on page 227.

Because the eastern limb dips at a shallower angle than the western one, the structure is termed an asymmetrical anticline.

Anticlines attract petroleum geologists the way a bulging sack of peanuts attracts chipmunks. The reason for this is clear-cut: one of the best places to look for deposits of oil and natural gas is where an anticlinal bulge traps them in a porous rock layer that is overlain by an impermeable one. Hence, the development of the Dupo oil field described above. This particular anticline, dear to the hearts of geologists and oil companies alike, appears to have been created by strong crustal forces near the boundary of the Mississippian and Pennsylvanian periods, and by smaller episodes of deformation before and after that time. This structure and others like it in Illinois probably are long-distance manifestations of major mountain-building events and continental collisions that took place a thousand miles to the east.

When you're done meditating on the subject of anticlinal geometry, continue up Imbs Station Road to the Y-intersection with Old Cement Hollow Road. Take Old Cement Hollow Road uphill and eastward. Soon you will be on the upland surface, where sinkholes, one of the classic surface expressions of karst topography, abound. They largely determine the way

On the western side of the Waterloo-Dupo anticline, St. Louis limestone beds dip steeply to the west.

227

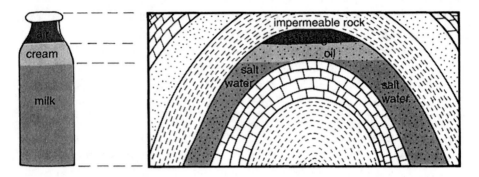

In the days before homogenization, the contents of dairy bottles could separate into milk, cream, and air. In the same way, crude-oil deposits trapped in an anticline separate by density: natural gas at the top, salt water at the bottom, and petroleum sandwiched between.

A sinkhole along the entrance road of Illinois Caverns State Natural Area.

road builders and farmers have used the land. Bear southward toward Columbia on connecting county roads. As you go, notice that some sinkholes drain well into the underlying rock, and others are temporarily plugged with debris and thus have partially filled with water. The sinkholes are formed by solution or by cavern-roof collapse. They often take the place of the small streams that would otherwise feed the major rivers. Instead of being confined to surface flow, the water is funneled into solution channels underground. The water usually reemerges: a stream suddenly rises apparently from nowhere; or, as at Falling Spring, it gushes out of a cliff face.

In the prim and tidy town of Columbia, merge onto Route 3 and head south past Waterloo to the Kaskaskia Road turnoff to Burksville. Signs there will guide you to one of the state's greatest geologic marvels, Illinois Caverns State Natural Area, which for many years was run as a commercial attraction until it was purchased by the state in 1985. This is the site where the spelunking gear is required. The access road to the parking lot passes a good example of a sinkhole. To descend into the cave complex, you must first sign a permit at the park office and demonstrate that you are properly equipped. The steep set of stairs leading down to the sinkhole opening to the cave proper would make a good backdrop for the final part of Berlioz's *Damnation of Faust*. There, among the dangling clumps of dripping moss, you get the feeling the hole will surely close up behind you, amid a wailing chorus of lost souls, as soon as you're inside. Instead, this is the entryway to a natural wonderland with immense appeal to geologists and nongeologists alike. From the entrance, you walk, crouch, and crawl along a subterranean stream channel that suggests latter-day vadose action is going on. The string of chambers has descriptively named sections: Cascade Canyon, Mushroom Passage, Marvin's Misery. This spectacular site, deep in the damp, ancient recesses of Mississippian limestone, reminds the visitor that geology is a three-dimensional subject.

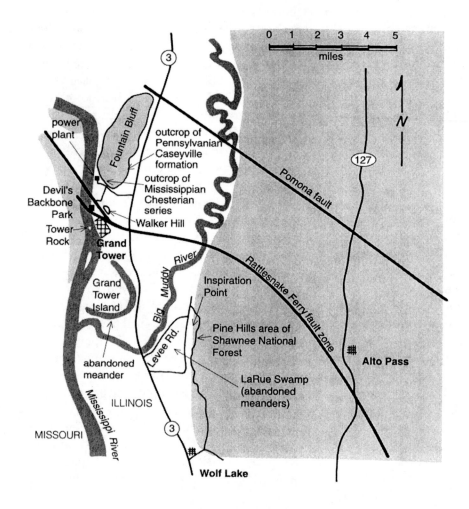

0 1 2 3 4 5
miles

N

3

power plant

Fountain Bluff

outcrop of
Pennsylvanian
Caseyville
formation

outcrop of
Mississippian
Chesterian
series

Walker Hill

Devil's
Backbone
Park

Tower
Rock

Grand
Tower

Grand
Tower
Island

Big Muddy River

Inspiration
Point

Pine Hills area of
Shawnee National
Forest

Levee Rd.

abandoned
meander

Mississippi River

ILLINOIS

MISSOURI

3

LaRue Swamp
(abandoned
meanders)

Pomona fault

127

Rattlesnake Ferry fault zone

Alto Pass

Wolf Lake

Southwestern Jackson County. The lightly shaded area is upland; the white area is bottomland. Note the Rattlesnake Ferry fault zone and the Pomona fault.

— 32 —
A Fault-Seeking Expedition
FOUNTAIN BLUFF TO THE PINE HILLS ESCARPMENT
Jackson County

Down by the village was the river, a whole mile broad,
and awful still and grand.
—Mark Twain, *The Adventures of Huckleberry Finn*

Jackson County, Illinois, is one of the outstanding places in North America where major themes of earth history unite with beautiful scenery and intriguing detail. If anyone you know needs to be persuaded how fascinating the different aspects of geology can be, bring that person hither. This big-river country, the land of Mark Twain, has a dreamy grandeur that accommodates and reconciles broad expanses of plain and swamp with soaring faces of unobscured rock.

The drowsy majesty of the place belies its dramatic geologic past. If you trace a diagonal line on a map between a point a little south of Alto Pass and one a little north of the town of Grand Tower, you will be delineating a feature called the Rattlesnake Ferry fault zone, a rupture in the earth's crust that is a component of the larger, 100-mile-long Ste. Genevieve fault system. Another part of the Ste. Genevieve system, the Pomona fault, lies a few miles north of the Rattlesnake Ferry structure. These faults are the product of large-scale stresses that originated on our continent's eastern flank. Recent mapping by the Illinois State Geological Survey's W. John Nelson indicates that the Pomona fault is the product of two spates of crustal movement. The first took place near the middle of the Devonian period, about 385 million years ago; the second, near the end of the Mississippian period, roughly 330 million years ago.

Faults can be remarkably complex and unique features. Specialists who chart and study them must have a good sense of three-dimensional geometry and the willingness to avoid rash generalizations. For instance, the geologists who described the Pomona fault discovered that it was originally a normal fault—one in which the side termed the hanging wall has been lowered in comparison to the side called the footwall. The second episode was typical of a reverse fault, where the hanging wall has moved upward relative to the footwall. The net result was that the Pomona's southwest side

231

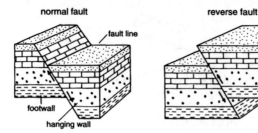

normal fault

reverse fault

fault line

footwall

hanging wall

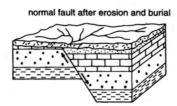

normal fault after erosion and burial

Normal and reverse faults. Subsequent erosion and deposition can make a fault trace invisible at the surface. —Illinois State Geological Survey

is slightly higher than the northeast side. The Rattlesnake Ferry fault zone has displacement that is much more impressive: the rock layers of the southwestern side—the hanging wall—have been thrust some 2,000 feet higher than the corresponding strata on the other flank.

One isolated example of how John Nelson and other structural geologists have determined a part of the chronology of the region's episodes of faulting illustrates the detective work involved. As in a mystery novel, small clues can sometimes lead to great jumps in understanding. At one place along the trace of the Rattlesnake Ferry feature, where the Mississippian and Pennsylvanian strata outcrop together, the contact between the rocks of the two periods is what is known as an angular unconformity. Here, the Mississippian beds dip a little more steeply than their overlying Pennsylvanian counterparts, suggesting that the Mississippian beds had been subjected to tilting and faulting even before the Pennsylvanian strata were laid down. The faulting continued into the Pennsylvanian, because the overlying beds of that age tilt, too, though not as steeply.

Unless you are an experienced field mapper, you probably won't find such direct evidence of the Rattlesnake Ferry fault zone; all too often, indirectness is the essence of geology. With a little patience you will find powerful if indirect indications of the fault's existence. Begin by making at least a partial circuit around one of Jackson County's great landmarks, the islandlike hill called Fountain Bluff. There is an unconscious wisdom in its name. It once was part of the main Mississippi River bluff, even though it now sits detached on the valley floor. Missourians have a right to regret the loss of a big chunk of real estate when they gaze at it across the river. It used to belong to their bluff. Apparently, it became detached during the

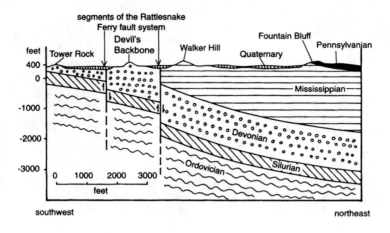

segments of the Rattlesnake
Ferry fault system

Simplified cross section showing the displacement of the earth's crust in the Devil's Backbone–Fountain Bluff area. Movement along the Rattlesnake Ferry fault zone has exposed Devonian limestone at Devil's Backbone and at Tower Rock. —Illinois State Geological Survey

Pleistocene, when the leading edge of the expansive Illinoian ice sheet blocked the Mississippi's channel to the east of Fountain Bluff. The glacial meltwater was forced to rapidly cut a more western sluiceway. When the ice retreated, the river stuck to its new route. Later, during the highest portions of the Wisconsinan Kankakee Torrent, Fountain Bluff must have temporarily been a true island, with water swirling by on all sides.

Fountain Bluff is made up primarily of Lower Pennsylvanian sandstones of the highly resistant Caseyville formation. This assemblage of clastic rocks accounts for many of southern Illinois's high points. It is especially visible at the base of Fountain Bluff's eastern side, along Route 3. On the southern end of the hill, though, along the road leading to the power plant, you can see a good exposure of the underlying Mississippian strata of the Chesterian series. And, when you head down the connecting blacktop to the town of Grand Tower, you'll be crossing over one main prong of the invisible Rattlesnake Ferry fault. It runs northwestward in the low area between Walker Hill and the Devil's Backbone. The other prong runs northward, too, along the riverfront side of the Devil's Backbone, which is now part of a town park superbly sited on the bank of the Mississippi.

The Devil's Backbone is what geomorphologists call a hogback, a narrow ridge of steeply inclined strata. These northeastward-dipping beds are suggestive of rock layers bent down by the drag effect of motion along a nearby fault. The hogback is composed not of Pennsylvanian rocks but of much

233

A large Mississippi River point bar west of the Devil's Backbone. The bridgelike structure in the background is a natural-gas pipeline.

older Devonian-period limestones of the Lingle and Grand Tower formations. These fossil-bearing beds of tiny, interlocking calcite crystals are gray to tan—a nice backdrop for the pale blue flowers of the cleft phlox plants that adorn them in early spring. At this intriguing site, try to comprehend the totality of the Rattlesnake Ferry fault zone. Millions of years' worth of erosion has reduced both its sides to a level surface. As a result, all the Pennsylvanian and Mississippian rocks on this upthrown side have been removed, which explains why only the much older, Devonian formations remain. On the northeastern side of the fault, in stark contrast, the Devonian strata are situated deep below ground.

Before you leave Devil's Backbone Park, walk down to the river's edge. Not far from the Missouri shore stands lonely Tower Rock, a resistant mass of Devonian limestone that was a landmark for Lewis and Clark, and a hazard to navigation for the riverboat skippers who eventually followed in their wake. President Ulysses S. Grant, no stranger to this region, declared Tower Rock a federal preserve long before our country's national park system came into being. If the water is not particularly high, you'll be able to venture out onto an exposed deposit of sand. This is a point bar: the low-energy side of the river, where sediments often accumulate in quantity because the water velocity is slower. See if you can find ripple marks that document how flowing water at a former, higher level molded the sand into miniature dunelike structures. Were you to dig a deep longitudinal trench into the point bar (and somehow keep water from filling it) you would also discover crossbedding patterns reminiscent of those in many of Illinois's ancient sedimentary rocks. Point bars may be visible one day and submerged the

234

next. As was convincingly demonstrated during the Great Flood of 1993—when houses even on the slope of the Devil's Backbone had flooded basements—the Father of the Waters will do whatever he wants, whenever he wants. Few other things in nature combine capriciousness and first-rate power so convincingly.

When you depart the Devil's Backbone, head eastward from Grand Tower until you reach Route 3. Take that road south. At the first left-hand turn south of the Big Muddy River, head east on the gravel road, which runs along the crest of one of the Mississippi Valley's many earthwork levees. In the summer of 1993, when I was snooping about in this locale, the Big Muddy had spread far over its banks. National Guardsmen posted here eyed my battered Chevy pickup with grave suspicion and warily waved me off. For all they knew, I might be one of the handful of farmers from farther south who tried to dynamite such barriers as this, to mitigate the flooding in their fields and towns downstream. But for every one of those potential saboteurs there were hundreds of local folks, assisted by volunteers from far and wide, who stoutly worked in the blazing sun to sandbag houses and the base of threatened levees, where water from the swollen river, under great pressure, bubbled up on the landward side. Geologists and ecologists can rightly question the long-term impact of levees that so naively tamper with an immense natural drainage system. But anyone who saw the plight of the river people in that time could only sympathize with their desperate and not always successful fight to stave off the immediate danger.

The levee road parallels the Big Muddy and gives you an unexcelled view of the bottomland formed by the Mississippi and this meandering, tree-lined tributary. In a few minutes' time, you'll come upon a 100-foot face of water-sculpted stone. This is the Pine Hills Escarpment, the gateway to the upland portion of one of the most ecologically precious parts of Shawnee National Forest. At the T-intersection under the cliff, ascend the escarpment by taking a left and then the first hard right. As you begin your climb up the bluff, you will pass a lowland parking area; continue on to the next pulloff, farther up. This is the upper terminus of the Inspiration Point Trail. A ten-minute hike and a minimum of labored breathing will bring you to one of the best vistas in the Midwest: the full expanse of the Mississippi plain, with the silent green-clotted waters of the great LaRue Swamp in the foreground. What a view! You stand atop a great exposure of the Lower Devonian Bailey limestone. Atop the Bailey is a separate formation, the extremely resistant Grassy Knob chert, composed of a rock that apparently was formed from the breakdown of silica-bearing hard parts of marine organisms. It has a characteristic orange-brown color. When split, it forms

The Pine Hills Escarpment.

LaRue Swamp occupies an abandoned river meander, far from the current channels of the Big Muddy and Mississippi Rivers.

the sharp edges and concave surfaces that geologists refer to as conchoidal fracture. These attributes were not lost on early native cultures of this region: chert was the preferred substance for arrowheads and other cutting, piercing, and scraping implements. One darker-toned version of chert is the familiar flint. Had Hollywood producers been a little more geologically generic, they would have named everyone's favorite Stone Age quarry worker Fred Chertstone instead.

Inspiration Point is a difficult place to pull oneself away from, especially in early spring when the redbuds, toothworts, and cleft phlox are in full bloom, and the dogwoods are about to be. Before general leaf-break is the best time to view the details of the great wetland below. You should be able to identify the scroll-like scars in the swamp that are abandoned river meanders from a time when the Mississippi cuddled close to this cliff.

When you leave, resume your southerly heading on the gravel road. It takes you through the rest of the heavily forested Pine Hills area. As you descend to low ground, take a good look at the beds of the small woodland streams you ford. They are floored—one could almost say paved—in Grassy Knob chert gravel. These small bits of rock, already some 400 million years old, may well survive longer than the rest of human history, as they continue their slow journey toward the sea.

View from Inspiration Point. The flat-bedded Bailey limestone stands in the foreground; beyond it lies the valley of the Big Muddy River.

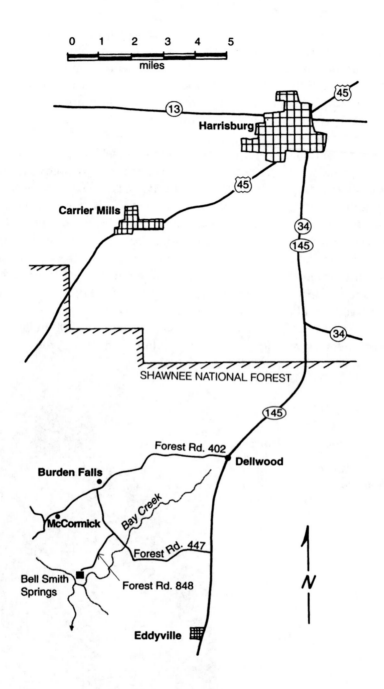

0 1 2 3 4 5
miles

45

13

Harrisburg

45

Carrier Mills

34
145

34

SHAWNEE NATIONAL FOREST

145

Forest Rd. 402   Dellwood

Burden Falls

Bay Creek

McCormick

Forest Rd. 447

Bell Smith
Springs   Forest Rd. 848

N

Eddyville

The easiest way to reach Bell Smith Springs is to take Illinois 145 south from Harrisburg, and then turn west on the unpaved Forest Road 447, just north of the small settlement of Eddyville. Signs will guide you the rest of the way. Park in the Bell Smith Springs parking lot at the end of the connecting Forest Road 848.

Sandstone Gorges, Rock Shelters, and a Natural Bridge
BELL SMITH SPRINGS
Pope County

Take almost any path you please, and ten to one it carries you down in a dale, and leaves you there by a pool in the stream. There is magic in it.
—Herman Melville, *Moby-Dick*

Bell Smith Springs, one of the state's great natural wonders, is the essence of seclusion. Situated at the end of a long gravel road that winds through a forested and largely unpopulated area, this site inspires a sense of time slowed almost to a standstill. To the botanist, it is a treasure chest stocked with more than seven hundred plant species; to the zoologist, it is a place where mountain lions once lived and where bobcats, timber rattlesnakes, and pileated woodpeckers still do. To the geologist and the lover of scenic locales, it is one of the most important parts of Illinois the glaciers never reached—a place where Bay Creek and its tributaries have carved a pattern of high cliffs and striking landforms in the massive sandstone bedrock.

Even on a busy day, when several carloads of hikers are present, the Bell Smith canyons maintain their sense of remoteness; people go their

Bell Smith Springs.

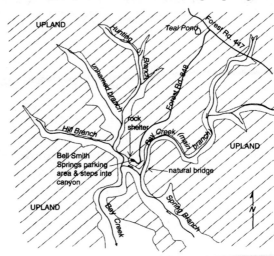

239

Rock shelters along the canyon wall at Bell Smith Springs. Note the detached and tilted block of Pounds sandstone that has fallen from the cliff face.

separate ways and soon disappear from sight. At the canyon trailhead, descend the long, narrow set of steps to the floor of the gorge. This magic stairway is the product not of some uncannily foresighted geologic process but of a United States Forest Service construction crew. Once at the bottom, you come upon an impressive piece of thoroughly natural architecture: a large recess in the cliff face that seems to have been expressly designed to protect visitors from passing rain showers. This feature and the others like it in southern Illinois are aptly called rock shelters. They are one of the characteristic landforms of this, the Shawnee Hills section of the Interior Low Plateaus Province.

As you examine the structure of this rock shelter, look for a spring issuing from the back wall. The water has its source in the nearby upland; it seeps down through joints and porous zones in the cliff-forming Pounds sandstone until it hits the nonporous Drury shale. Both the Pounds and Drury strata are subsections, or members, of the 310-million-year-old, Lower Pennsylvanian Caseyville formation. When the water's downward progress is checked by the Drury shale, it moves sideward until it emerges into the open air. Over thousands of years, this downward and outward motion of the seepage, together with the associated freeze-thaw action of the colder months, has dislodged sections of rock to produce the hollow and overhang.

The upper surface of the natural bridge at Bell Smith Springs.

Up and down these canyons you can find other rock shelters. Near them stand large detached blocks of stone, both upright and toppled, that were once a part of the solid cliff face.

Follow the trail signs for the site's number-one attraction, the natural bridge. If you visit during the spring freshet or at other times of high water, you will probably find yourself fording a swollen Bay Creek, when high-topped ooze shoes or expendable old sneakers are a must. Be careful not to lose your footing on the slippery stones of the streambed. Six inches of soaked trouser legs is a small price to pay to behold this famous landmark. The natural bridge, about 150 feet long and 60 feet tall, is the largest such structure in the state. It can be approached directly from its base or in a more circuitous way via the trail that snakes around and up the far bluff until it leads straight across the bridge's ramplike upper surface. To quickly get down to the base again, you can descend on a series of exposed iron rungs anchored in the vertical face; however, these are not recommended for anyone prone to vertigo or acrophobia, and the rungs can be slick. People with nongrip soles are well advised to take the long way back.

The origin of this marvelous landform is similar to that of the rock shelters. You could say it is a rock shelter taken one step further. Long ago, this was part of the unbroken cliff of Pounds sandstone. A rock shelter then

241

Crossbedded sandstone at the rear of the natural bridge.

formed in the manner already described. Some distance back from the front edge there was a joint, a vertical fissure along which there was no sizeable displacement of the strata. In time, water pouring down through the joint enlarged not only the shelter at the base but also the fissure. This process continued until the rock behind the present bridge was completely pried and eroded away.

A close examination of the rock face at the back of the bridge reveals a distinctive pattern called crossbedding. This elegant set of curving traces indicates that the sediments forming the Pounds sandstone were not deposited in still waters but came from an environment where current action was strong. Extensive study of this member has revealed that the rock represents alluvial deposits laid down by a Pennsylvanian river system flowing south or southwest from higher ground to the north. If you inspect a clean, unweathered sample of the Pounds sandstone, you will find it is composed of relatively coarse, white quartz particles, with little or no mica and other minerals present. Pebble-sized grains are not uncommon. The stream capable of transporting these heavier sediments must have been fast flowing.

As you explore other parts of this lovely maze of canyons, you will find other distinguishing features of the Pounds sandstone. In early spring, before leaf-break—as always, a much better time than midsummer to see geologic detail—the light tone of the cliffs gives way in places to a reddish brown that marks the zones where water spilling over the top has wetted the rock. If you are an aficionado of auditory delights, you'll also enjoy the gradation of sounds—everything from concussive slapping effects to a rarified tinkling—created by the falling sheets and cascades of water. The mist that may wet you a bit is more refreshing than irritating.

Even the sandstone used as steps and flagging at points along the trail has a tale to tell. Look for petrified ripple marks, another sign of the active, shallow-water conditions in which these ancient sands were laid down. You may spot a more perplexing feature on the Bay Creek wall face, a little downstream from the entry stairway. A small, solitary boulder sits there incongruously, embedded in the vertical expanse of much finer sediments. Here's your chance to do a little creative geological speculation. Try to come up with a plausible theory of how this large rock came to be trapped here. Do you think the boulder was of local origin, or was it carried many miles before it was laid to rest?

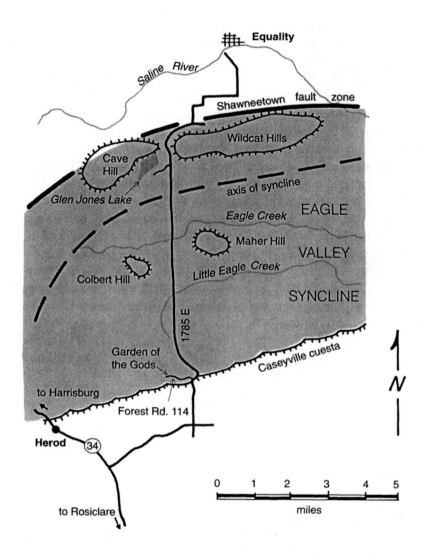

The simplest way to reach the Garden of the Gods is to take Illinois 34 south from Herod. About 2 miles south of town, you'll see the first of several signs directing you to the Garden of the Gods parking lot. The Eagle Valley syncline is shaded gray.

~ 34 ~

What Goes Down Must Come Up

THE GARDEN OF THE GODS AND
THE EAGLE VALLEY SYNCLINE
Saline and Gallatin Counties

*We find these strata that were originally formed continuous in their
substance, and horizontal in their position, now broken, bended, and
inclined, in every manner and degree; we must give some reason in
our theory for such a general state and disposition of things; and we
must tell by what power this event . . . had been brought about.*
—James Hutton, *Theory of the Earth*, 1795

Any traveler who has visited the southernmost reaches of Illinois knows
it as a land apart, where the state's stereotypical scenes of flat farmland and
sprawling urban centers do not pertain. On entering the unglaciated Shawnee
Hills area, you discover that driving has become a three-dimensional affair.
The rolling landscape is lovelier, the people are more gracious, and the sunlight, even on the hottest summer's day, seems gentler. But this image of a
peaceable kingdom belies the fact that Illinois's southern tip, compared with
the rest of the state, has had a violent geologic past. Faults and large-scale
fold structures crisscross the region in numbers that must have amazed the
first geologists who discovered them. If any portion of the Prairie State is
a first-class geologic jigsaw puzzle, this is it.

Only in the past few decades, with the development of sophisticated
remote-sensing and mapping techniques, have geophysicists and other earth
scientists pieced together a fairly complete picture of how this area's complex geology came to be. The first chapter of the story is the most surprising: it tells not of a solid and stable Midwest situated far from the various
complications of plate tectonics but of a place subjected to the full brunt
of the forces that break and build continents. In late Precambrian time or
early in the Cambrian period—long before the oldest surviving sedimentary rocks were deposited on top of the ancient igneous rocks of the basement complex—this section of North America began to split apart, along
a southwestward-trending structure that goes by the unwieldy composite
name of the Rough Creek graben-Reelfoot rift. A graben is a long, troughlike
fault complex in which the central area has been dropped down relative to

245

the sides. One can see grabens in the Rio Grande Valley of northern New Mexico, in Iceland, in East Africa, and in other locales the world over.

An examination of our planet's long record of plate-tectonics activity shows that some giant rift systems remain active until their two sides are completely detached, like the huge Atlantic Ocean rift that now separates the New World from the Old. Still, other rift zones quit working before the landmass in question is completely sundered. Such was the case with the Midwest of late Precambrian or early Cambrian time, when a super-continent apparently was suffering the throes of more complete dismemberment elsewhere. As it turned out, the Rough Creek-Reelfoot structure did not succeed in unzipping the Midwest from the Southeast.

While this aborted continental breakup was completed by the end of the Paleozoic's first period, the Cambrian, its legacy lived on in one sense. It had created a persistent zone of weakness in the earth's crust. But later, other factors would come into play. Late in the Devonian period, about 385 million years ago, the old faults were reactivated by new stresses. And closer to the end of the Paleozoic, probably near the dividing line between the Pennsylvanian and Permian periods some 280 million years ago, southern Illinois's now-massive blanket of sedimentary strata was subjected to a disturbance that may have been linked to the continental collision and mountain-building that was under way a thousand miles to the east. The compressional stresses created by that great chain of events reactivated the ancient Precambrian-Cambrian faults. Later, probably during the Early Jurassic breakup of the supercontinent Pangaea about 200 million years ago, there was still more movement. This time around, though, a general pulling apart, rather than compression, was the rule. One of the most important features formed in this long sequence of crustal strains was the Shawneetown fault zone, which runs from western Kentucky through Illinois's Gallatin, Saline, and Pope Counties. It is an excellent indicator of the long-lasting, complex tug-of-war the area experienced. In the Precambrian-Cambrian phase of rifting, the Shawneetown structure behaved as a normal fault; during the Pennsylvanian-Permian phase, as a reverse fault with its southern side thrust upward; and during the Jurassic, as a normal fault once again, with the southern side lowered to a level below its original position.

Even in historic times, sporadic earthquakes have served as reminders of this region's eventful past. The famous New Madrid shocks of 1811–1812, probably linked to the old Reelfoot rift, included some of the strongest tremors ever recorded in North America. One eminent seismologist has given them a maximum Richter Scale rating of 8.7—compared with the 7.9 rating assigned to the great San Francisco earthquake of 1906. More recently, the most powerful earthquake centered in Illinois—at Richter 5.5—hit Hamilton County in 1968. Current-day scientists, searching for telltale earthquake features in unconsolidated sediments on Illinois riverbanks, have uncovered evidence that tremors have been more frequent in this state than we supposed.

Of all the places in southernmost Illinois that demonstrate the area's complex geologic makeup, the Garden of the Gods in Shawnee National Forest may be the most breathtaking. The overlook is at the end of a short trail. There you can take in the broad expanse of wilderness scenery from the crest of the Caseyville cuesta, one of the state's great geographical features. A cuesta (pronounced *KWEH-sta*) is a ridge that is asymmetrical in cross section: one of its slopes is gentle; the other, which resembles a cliff face, is steep but usually not completely vertical. The rock that takes on such fantastic forms here is the Pounds sandstone member of the Lower Pennsylvanian Caseyville formation. This is the same bed exposed in a different setting farther west, at Bell Smith Springs (see the preceding essay).

Some features here—the vertical joints and the crossbedding patterns typical of sedimentary deposits laid down in agitated water—are common sights elsewhere, too. But what is unique here is the prevalence of weird, wavy bands that cut right through the crossbedding traces on the sandstone face. These bands are Liesegang rings—concentric zones of concentrated

The Caseyville cuesta, the high, forested ridgeline, from the south.

The Garden of the Gods. The enduring Pennsylvanian Caseyville sandstone forms rounded, heavily jointed masses.

Liesegang rings in the Caseyville sandstone. Dime for scale.

iron oxide that came into being long after the sediments were laid down. Because these hard, well-cemented rings erode more slowly than the rest of the sandstone, they stand out prominently.

The view westward from the overlook reveals a hilly, forested terrain as far as the eye can reach. A rural New Englander would feel right at home, were it not for the sultry summers and the lack of white pines and stony glacial till. To the north, you can get a peek at one end of this region's most geologically fascinating low areas, the valley of Eagle Creek. This stream course runs parallel to the axis of a major syncline that shares its name. A syncline is a series of rock beds that have been bent downward toward the middle. At the Garden of the Gods, you are on the southern side of the Eagle Valley syncline; accordingly, the Pounds sandstone here dips gently northward. To make a transit of this intriguing structure of downwarped Pennsylvanian strata, depart the Garden of the Gods and head back down the hill. Instead of returning to Route 34, however, turn north on the paved Forest Road (1785 East).

The first leg of the transit takes you over high ground held up by the tough Grindstaff and Caseyville sandstones. When you descend into the lowland of Little Eagle Creek, you are on top of younger, less resistant, coal-bearing strata. The next upland area, best delineated by Colbert Hill to the west and by Maher Hill to the east, is made up of beds capped by the

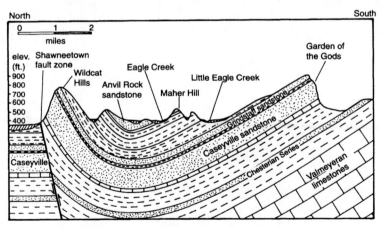

Simplified cross section of the Eagle Valley syncline, traversed north-south by 1785 East. The vertical scale is exaggerated for clarity. —Illinois State Geological Survey

relatively resistant Anvil Rock sandstone—the youngest layer in the entire syncline. So far, all these rock beds have had a gentle northward dip. But about a mile and a half after you cross the main branch of Eagle Creek, the beds start to tilt more deeply, and in the opposite, southward direction. You have just crossed the syncline's axis. Now the rock beds at the surface are in reverse order: what went down earlier on is coming up again. When you reach the left-hand turnoff for the Saline County Fish and Wildlife Area, you are on the western edge of the Wildcat Hills. They are a mirror image of the Garden of the Gods ridgeline; and they too are formed from the older, more resistant Grindstaff and Caseyville sandstones. Despite the similarity, as you pass over the east-west Shawneetown fault zone, there are also detached, downward-pointing wedges of much older, Devonian rock. These wedges are fragments of the crust that have failed to keep pace with the up-and-down movements of hundreds of millions of years.

If you continue north and east on the connecting county roads, you will encounter a striking change in the landscape and in the subsurface geology as well. Just north of the Wildcat Hills and the Shawneetown fault zone, the syncline comes to an abrupt boundary where it meets the southern fringe of the expansive Saline River valley. Besides being its own structural feature, the Eagle Valley syncline is part of another—it is the Shawneetown fault's southern, upthrust block. North of the fault line, the strata belong to the same formations that you've already passed over, but they have a completely different configuration. As the uniform flatness of the terrain suggests, they are not tilted or downwarped, but horizontal. This is one more demonstration that the geologic makeup of Illinois's southern tip changes quickly and often.

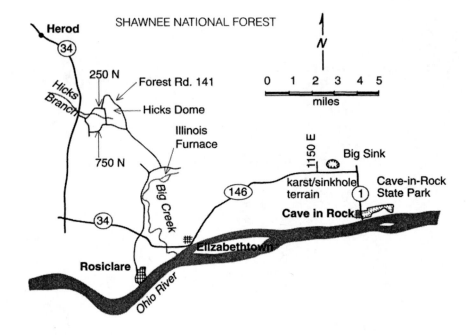

SHAWNEE NATIONAL FOREST

Hicks Dome is most easily reached by taking Illinois 34 south from Herod. Four miles south of Herod, you will cross the stream of Hicks Branch. Turn eastward onto the gravel road that parallels its south bank. This county lane winds about until it straightens and becomes 250 North. Here you first see the mound-shaped profile of Hicks Dome in front of you.

To reach the Illinois Furnace, you can test your navigational skills on the hilly county roads heading south from Hicks Dome, or you can resume your southward journey on Illinois 34. At the Rosiclare junction, turn north and follow the signs.

— 35 —
The Geologic Wonders of an (Almost) Undiscovered Country
HICKS DOME TO CAVE-IN-ROCK STATE PARK
Hardin County

Underneath the grass roots begins the silent, immobile
kingdom of the minerals.
—H. P. Whitlock, *The Story of Minerals,* 1925

To the person blessed with the opportunity to travel the length and breadth of Illinois, there is no more happy discovery than Hardin County. Still, the place is a paradox. It contains some of the state's most beautiful scenery, including untrammeled upland wilderness and breathtaking views of the stately Ohio River. The people here—who speak mainly in an unhurried Southern dialect—are some of the most hospitable, interesting, and unabrasive folks in the nation. Despite that, the region retains an at-mosphere of remoteness. For the sake of the local economy, we might wish it were better known to tourists; much of its appeal lies in the fact that it isn't. At any rate, people interested in the earth sciences have a special reason to make the pilgrimage to this place of many geologic wonders. It is here that the most attractive member of Illinois's mineral kingdom, fluorspar, has formed amid the shattered rock beds of the region's fault zones.

As noted in the preceding essay, Illinois's far-southern section has been a focal point for large-scale deformation ever since the end of Precambrian time, approximately 600 million years ago. The creation of a major rift system, probably linked to the breakup of a Precambrian supercontinent, was only the first episode of several geologic upheavals here. For example, many of Hardin County's unusual landforms, subsurface structures, and economi-cally important mineral deposits came into being, directly or indirectly, in response to violent crustal activity late in the Paleozoic era.

This essay's first point of interest, the great oval structure of Hicks Dome, is also its most mysterious. With the help of a topographic map or a state atlas, you can circumnavigate the whole structure if you wish. One of the best vantage points is along the roadside due north of the dome.

253

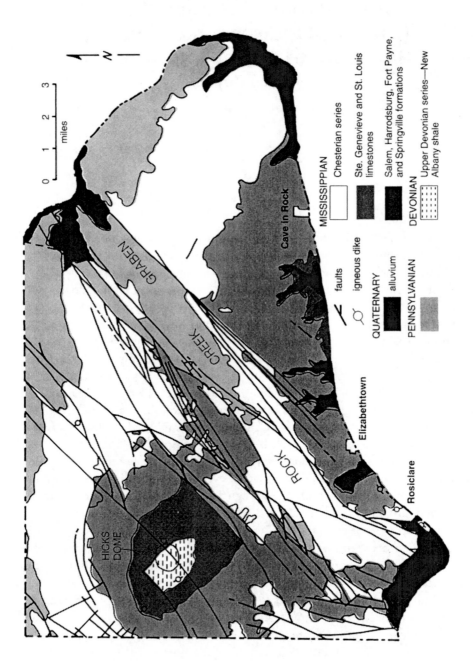

MISSISSIPPIAN

☐ Chesterian series

▓ Ste. Genevieve and St. Louis limestones

▨ Salem, Harrodsburg, Fort Payne, and Springville formations

DEVONIAN

▦ Upper Devonian series—New Albany shale

faults

⚬ igneous dike

QUATERNARY

■ alluvium

PENNSYLVANIAN

▒

Cave in Rock

Elizabethtown

Rosiclare

HICKS DOME

GRABEN

CREEK

ROCK

N

0 1 2 3
miles

Simplified geologic map of Hardin County. —Illinois State Geological Survey

Hicks Dome viewed from the north. The wooded crest of the dome is formed of Devonian limestones; the lower, cleared flank is underlain by the Devonian New Albany shale.

The central upland portion of Hicks Dome looks much like the county's other hills—or at least it does to the untrained eye. But a geologic map reveals how strange it really is. For it is composed not just of the Mississippian carbonate-rock formations common elsewhere in this part of the county but also of strata from the preceding geologic period, the Devonian. As you drive about, you may not find direct evidence of this in the form of outcrops; but the hill's vegetation patterns and overall profile provide some helpful hints. The dome is capped with resistant, Early and Middle Devonian cherty limestones that include the Lingle and Grand Tower formations seen and discussed in essay 32. This highest section is still covered with woodland stands. The surrounding low area rests on the Upper Devonian New Albany shale—the same formation that forms the bed of Lake Michigan. Shale, often an easily eroded substance, usually expresses itself in low ground and gentle slopes. That is definitely the case here. Note how the surface has been cleared for agriculture. Still farther from the center stands a forested outer rim of Lower Mississippian limestone and chert of the Fort Payne formation. If you could behold a cross-sectional slice of the dome, you would see that its strata dip downward and away from the center

255

on all sides. In addition, the surface drainage pattern is unusual: the rills run down its sides to form a radial pattern, like spokes from the hub of a wheel.

Geologists have determined that the Devonian core of Hicks Dome has been displaced an almost unbelievable 4,000 feet upward, relative to the rocks of the surrounding area. What could possibly have produced such an effect in this solitary locale? It is as though a titanic thumb has poked the fragile crust from far below and succeeded in popping the Devonian beds far above their assigned resting place. One leading theory suggests that this giant thumb was a mass of magma, rising from deep in the earth's interior. Hot gases accumulating at the top of the magma apparently found their way to the surface, not in a prolonged period of gradual leakage but in one or more explosive episodes. Ample evidence of this catastrophe exists in the form of breccia (rock composed of shattered angular fragments), which has been found in selected locales near the heart of the dome. In addition, many dikes—narrow sheetlike bodies of intrusive igneous rock—have sliced upward through the sedimentary strata.

Dikes appear elsewhere in Hardin County and surrounding areas. Unfortunately, they are difficult to locate, because their outcrops weather and disintegrate rapidly in this moist, warm climate. According to Eric Livingston, chief geologist of the Ozark-Mahoning Company, the only really obvious dike structure in this part of southern Illinois outcrops on the bank of the Ohio River, just upstream from Rosiclare. To see it, though, you'll have to rent a boat. Though you may not have the opportunity to scrutinize this rare igneous rock firsthand, at least you'll know that Illinois is not 100 percent sedimentary, as many people suppose.

The next locale on this itinerary is the Illinois Furnace site, a pleasant treesy area along the gurgling waters of Big Creek. Illinois Furnace is a stone-clad memento of the time more than a century and a half ago when Hardin County was a producer of iron, as well as of fluorspar. The furnace you see here is an attempt at a reconstructed version of the original structure. The iron was extracted from limonite, a yellowish or orange-tinted mineral that was found in rather skimpy quantities in the local Mississippian limestones. The Iron Age of Hardin County did not last long; its industry was eclipsed by large-scale fluorspar production and also by galena mining, which lasted into the second half of the nineteenth century, even though it never produced an output comparable to that of Jo Daviess County (see essay 1). An informative Forest Service display, located in front of the furnace, explains the historical significance of the site in detail.

When you're ready to push on, you may wish to take a turn through the river town of Rosiclare, where the Ozark-Mahoning Company is

The reconstructed
Illinois Furnace.

headquartered. For quite some time, Ozark-Mahoning was the state's only surviving major producer of fluorspar, but in December 1995 it announced that its last deposits, in the Cave in Rock district, had been mined out. The mining and ore processing have ceased, but you can inspect and purchase fluorspar specimens in rock shops in Rosiclare and neighboring communities. There are also plans (and more significantly, funding) to open a fluorspar museum. It will feature exhibits devoted to this most important mineral, and to this most important industry, of Hardin County. Check with the folks in Rosiclare to see if the museum has opened for business.

In this area, fluorspar—known to mineralogists as fluorite and to chemists as calcium fluoride—owed its existence to the same late Paleozoic tectonic activity that produced many of the region's faults. In all likelihood, the once-extensive Hardin County deposits formed when fluorine-rich solutions moved upward from a magma body emplaced during the Permian period. These solutions reacted with the host Mississippian limestones to form horizontal and often extensive bedded deposits. Additionally, they precipitated into irregular vein deposits situated in more or less vertical fault fissures. One particularly rich ore zone is associated with the Rock Creek graben, a linear fault complex that runs across the county from the northeast

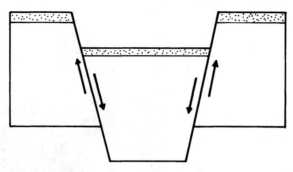

Idealized cross section of a graben. The block of the earth's crust in the center has dropped along the two faults that flank it. —Illinois State Geological Survey

to the southwest. Pennsylvanian sandstones of the Caseyville formation form the heights in the center of the graben. Ironically, they are part of the middle, dropped block; but because they are more resistant to erosion they now stand above the surrounding Mississippian carbonate strata. The last of the Ozark-Mahoning mines were located at great depths—more than a thousand feet down—on either side of the graben. They were of the room-and-pillar design that is also used by many of Illinois's larger underground coal mines.

Minerals often have more than one form. Native sulfur can be crystalline or amorphous, and the various types of quartz come in a rainbow of colors. Fluorspar exhibits a similar penchant for variation. It has a characteristic glassy luster, but it can be colorless, white, green, yellow, purple, pink, or blue. Few people who have seen a display of its contrasting forms would quibble with fluorspar's official status as the Illinois state mineral.

In the final years of mining, Ozark-Mahoning faced stiff competition from fluorspar sources in such places as Mexico, China, and South Africa. Accordingly, it chose to specialize in the production of acid-grade fluorspar—which is required to be at least 97 percent pure after processing. This high-quality product is the source of hydrofluoric acid, a chemical used in refrigerants, fuels, plastics, and other substances upon which modern civilization depends. As a sideline, the company also extracted barite and zinc found in association with the fluorspar, and marketed them as ingredients in several compounds, including the special lubricating mud used in drilling oil wells. (Other economically significant minerals found in the Fluorspar district included copper and lead.)

An entirely different sort of geologic wonder—in this case, gloriously exposed to public view—lies farther east. Depart Rosiclare on Route 146 toward Elizabethtown. Not long after you pass that community, you will start noticing that the landscape is dotted with circular depressions, some of which are filled with water. If you have been to the Waterloo area of Monroe County, you'll have no trouble identifying these features. They are sinkholes, key indicators of the unique limestone solution-terrain known as karst topography. Because these funnel-like hollows are so prevalent here, I propose that the state legislature redesignate this stretch of road the Karst Memorial Highway. It would be fitting, since the grandest of Illinois sink-holes, named simply enough Big Sink, is located a little to the northwest of the Route 146 junction with Route 1.

On reaching that intersection, turn south on Route 1, enter the nonhyphenated town of Cave in Rock, and follow the signs to the hyphen-ated Cave-in-Rock State Park. If you are going to see flocks of people anywhere in Hardin County, this is the place. The park's chief attraction—besides the sweeping Ohio River vista—is the massive riverfront cave. Up until the 1830s, this impressive natural shelter was the lair of desperadoes who plundered unwary settlers making their way down the big waterway. The imposing rock bed exposed here is the Mississippian St. Louis formation, which is also a familiar sight on the eastern cliffs of the Mississippi River (see

Dark, oval chert nodules in the jointed St. Louis limestone near the Cave-in-Rock entrance.

The entrance to Cave-in-Rock.

essays 20 and 31). As you walk around the bluff face to get to the cave entrance, inspect this light gray stone carefully. It is chock full of chert nodules, which resemble giant raisins embedded in an unchewable cake.

For a long time, Cave-in-Rock has moved geologists to theorize. The consensus is that it was once a much longer cavity; over the course of many thousands of years, the Ohio has swept away its southern portion. Interestingly, it appears that the cavern is not directly linked with the sinkhole complex to the north—it is much older, with its origin in late Tertiary time. The sinkholes probably came into being later, during the wet conditions of the Pleistocene epoch. The Ice Age had an impact even where its glaciers never reached.

As you'll see when you look at the cave's floor, there is an empty channel, suggesting that a subterranean creek once flowed here. This might tempt you to conclude that Cave-in-Rock is a vadose feature—one created by an underground stream that carved this cavity above the local water table. However, the cavern angles downward and out of sight, beyond what appears to be its northern end. This means the water that created it must have moved uphill, at least in one section, to reach its final destination, the Ohio River. Since it is unlikely that the laws of gravity were suspended at any point in geologic history, even in magical Hardin County, one must assume the water was forced upward by substantial hydrostatic pressure. Consequently, Cave-in-Rock must have been largely formed in phreatic conditions, where the eroding agent was slow-moving groundwater rather than a conventional flowing stream.

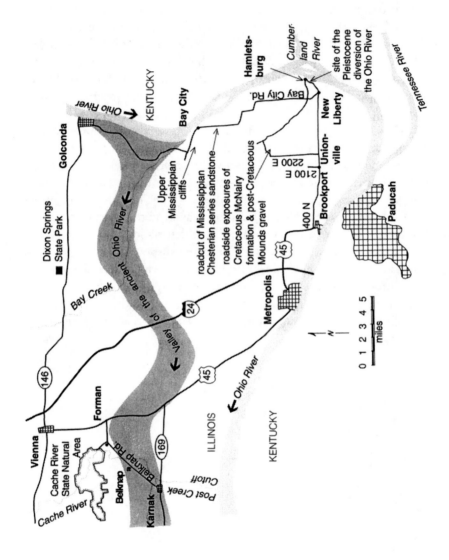

The Ohio River country in southern Illinois.

The Little River That Couldn't
HAMLETSBURG TO CACHE RIVER STATE NATURAL AREA
Pope and Johnson Counties

If a river takes another course, made by art, or nature, or some
violent cause . . . it leaves its former channel dry.
—Bernhard Varenius (1622–1650)

Every resident of northern Illinois should be required to pay at least one lingering visit to the Prairie State's Ohio River country. The many attractions of Hardin County, extolled in the previous essay, are one excellent reason to do so; but that is not the sum of the region's appeal. Farther south and west lies an even less frequented area, the peaceful land of Pope County. The route between Hamletsburg and Golconda, and farther westward, is so bucolic and even otherworldly that you may forget you've ever seen a strip mall or a fast-food restaurant. It is an excellent place to be deprogrammed.

The tiny town of Hamletsburg, the starting point of this exploration, occupies a surprisingly significant site. The ancient Ohio River apparently succeeded in changing its course by spilling over a low divide to occupy its present valley downstream. This momentous event, the permanent diversion of one of the continent's greatest waterways, probably occurred late in the Pleistocene epoch, roughly 15,500 years ago. A vast volume of glacial meltwater, the Kankakee Torrent, filled the stream courses of the lower Midwest with an immense amount of glacial outwash. In some places, preexisting valleys were buried with 150 feet or more of sediments washed down from the retreating lobes of the Wisconsinan ice sheet. It is difficult to overestimate the impact that the relatively sudden arrival of so much material had on the drainage system of North America's interior. In addition to changing the course of the Ohio, it forced the even larger Mississippi River to abandon its channel southwest of Cape Girardeau, Missouri, and create a more eastern channel that passes by the town of Thebes, Illinois.

When you arrive in Hamletsburg, drive along the blacktop closest to the waterfront and take a good look at the modern Ohio River at the beginning of the youngest section of its thousand-mile-long valley. To your east, behind two midchannel islands, the fabled Cumberland River

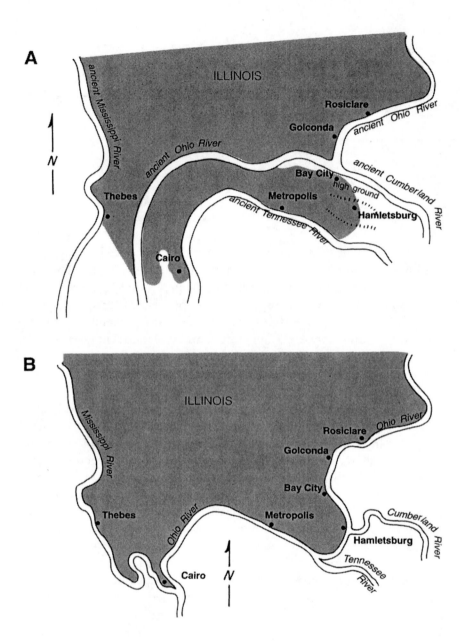

A: *The most widely accepted interpretation of the ancient river courses at Illinois's southern tip.* **B:** *The rivers' modern valleys.* —Illinois State Geological Survey

joins the Ohio after passing through the Tennessee Valley Authority's dammed and flooded Land Between the Lakes region. Geomorphologists who have studied the evolution of America's eastern rivers believe that the Cumberland—unlike the Ohio and the Mississippi—is an extremely venerable stream. It may have wended its way through the coal-swamp forests of late Paleozoic time and carried its sediments toward long-vanished inland seas, all before the dinosaurs came to be, and before North America broke away from the supercontinent of Pangaea to form the Atlantic Ocean. But according to the most widely accepted interpretation of its recent history, the Cumberland now meets the Ohio in a different place than it did before the Pleistocene. The facts suggest that its confluence was farther north, near Bay City.

You can gather your own supporting evidence for this theory of regional stream diversion. Head up the inland blacktop that is the main thoroughfare between Hamletsburg and points north. This road cuts across the contact between unconsolidated sediments of Cretaceous age and the much, much older sandstones of the Upper Mississippian Chesterian series. (If you'd like to make a short detour to examine roadcuts of the Cretaceous McNairy formation, proceed west on Hamletsburg Road to its junction with 2200 East, about 1 mile east of Unionville. Turn north on the latter, unpaved road, and continue on it for about 4 miles. You will see exposures of white and red-stained sand. That's the McNairy. In the highest places, where the hilly track bends eastward, you may even find that it is overlain by the much younger, post-Cretaceous Mounds gravel, a deposit whose enigmatic status and origin are described more fully in the following essay. Distinguish the Mounds from the lily-white McNairy sediments by its reddish color and by the brown chert pebbles it contains.)

When you reach Bay City, continue north toward Golconda. Soon you cross Bay Creek, the same stream that cut some of the canyons of Bell Smith Springs far upstream (see essay 33). Pull over for a moment to scan the cliffs of Upper Mississippian strata on the other side of the water. Before the Kankakee Torrent, this imposing wall of rock probably loomed not over this lazy fisherman's creek but over the broad meeting place of the ancient Ohio and the ancient Cumberland. Instead of continuing south as it does now, the ancient Ohio rolled westward through the expansive valley you see here. Only when that valley became clogged with Wisconsinan outwash did the water back up, swell southward, and eventually breach the Hamletsburg divide.

Despite this decisive change in the river's course, the great valley of the ancient Ohio remains. Here at its eastern end, the channel is now occupied by Bay Creek; farther west, by the westward-flowing Cache River. These

Bay Creek near its confluence with the modern Ohio River. Mississippian strata form the bluff in the background. Many geologists believe that the ancient Ohio and ancient Cumberland Rivers met here, before a major regional stream diversion late in the Pleistocene epoch.

two humble successors of the ancient Ohio are textbook examples of underfit streams—waterways that are too small to have carved the great path they follow. You will see more of the abandoned ancient Ohio Valley in a while; now, however, cross its floor heading north and rise up onto the northern bluff, on the road to Golconda. This sun-drenched upland surface, brought to you courtesy of the resistant Mississippian bedrock, is a joy to drive on.

When you reach the handsome river town of Golconda, bear west on Route 146. You are now traversing more hilly terrain, where the crust of the earth has suffered much deformation and faulting. Don't pass by Dixon Springs State Park without taking a look at its crossbedded and naturally sculpted masses of Pennsylvanian Caseyville formation sandstones. This shady retreat is situated in the central, downthrown block of a fault complex that is almost an identical twin of Hardin County's Rock Creek graben.

Next, resume your westward trek on Route 146 toward Vienna. At that town, turn south onto Route 45. The road descends from the Mississippian highland to the ancient Ohio Valley and the Cretaceous lowland. After about 9 miles, turn west on Route 169. Welcome to the region known as Little Egypt. Theories abound on the subject of how this area came to have so

266

many towns with Egyptian monikers; perhaps early American settlers thought this lowland was America's answer to the floodplain of the Nile.

On arriving in Karnak, you will not find the ruins of ancient pharaonic palaces but rather a sleepy railroad town inhabited by very friendly people. At the main intersection, turn north onto Belknap Road and head across the tracks and the one-lane bridge. Here you cross the swampy terrain of the Cache River. Believe it or not, you are traversing the northernmost inland part of America's Coastal Plain Province. This lowland extends from Massachusetts along the seacoast to Florida, and around the Gulf of Mexico coast to southern Texas. It also extends up the lower Mississippi Valley as far as Illinois's southern tip. In many ways, both the geology and the plant life of this area have more in common with western Tennessee and Louisiana than they do with the rest of Illinois.

As you head up to Belknap, take a look at the broad plain surrounding you. It is drained by the Cache, the little river that couldn't have created this great channel, given its weak and sluggish nature. You are still driving across the floor of the ancient Ohio. Were you able to see the great river as it was thousands of years ago, you'd find that it continued on until it met the ancient Mississippi well to the west of their present junction at Cairo.

The Cache is so thoroughly underfit and unambitious that it has trouble draining this area to the satisfaction of the local farmers. To speed up its flow, they have straightened it in several places. They have also drastically shortened its course to the Ohio by constructing a long cutoff canal. Such stream-course modifications are called ditching or channelization. Ditching

The abandoned floodplain of the ancient Ohio River, northeast of Karnak. One small stream, the Cache River, flows through this valley in modern times.

This view from the pedestrian bridge at Cache River State Natural Area looks upstream at the junction of the Cache River and Dutchman Creek (right). Ditching and cutoff canals downstream have caused the streams to cut deeply into the banks.

Bald cypresses rise from the still surface of Heron Pond Swamp.

is a time-honored way of draining waterlogged agricultural land; it also often triggers severe environmental problems. To see why—and to take one of the most instructive and soul-restoring hikes anywhere in the state—continue northeast past Belknap, follow the signs to the Cache River State Natural Area, and park in the terminus lot at the end of the long gravel access road. Before you get there, though, note a good example of human-induced erosion, where many decades' worth of vehicular traffic has worn the roadbed below the level of the surrounding ground.

This State Natural Area would be better termed a Snake Natural Area. Because this is a superb piece of Coastal Plain swampland, you may encounter some of the habitat's many reptilian residents. Fortunately, these unjustly maligned creatures have no desire to have *Homo sapiens* for lunch. If you steer a wide berth, so will they. Don't be deterred, but do be observant.

The first point of interest along the trail is the view from the pedestrian suspension bridge over the Cache River. Looking upstream, you will see where the Cache is joined on the right by its Dutchman Creek tributary. Note how these two streams, originally so torpid and incapable of erosion, have cut down deeply to reveal steep banks. This is the direct result of twentieth-century ditching and canal construction downstream. The water in shorter, straighter streams moves at a greater velocity because the regional gradient has been steepened. The result is a rejuvenated drainage net with a dramatically increased ability to move and remove its area's sediments.

As you head farther down the trail paralleling the Cache, you will see why state conservation officials are deeply worried about the alarming rate of erosion. On your right, the river is now incised approximately 8 feet lower than its floodplain. On your left, almost at the level of your boot soles, stretch the still waters of Heron Pond, the magnificent bald cypress swamp that has flourished over the centuries in a discarded channel of the Cache. It is one of the chief glories of Illinois's wilderness preserves. In places, the swamp water has cut steep drainage rills across the trail and down to the river. If the swamp succeeds in draining away more thoroughly, this priceless natural setting and its unique community of plants and animals will be lost forever. State conservation officials plan to install two weirs in the river to raise its level and slow the rate of downcutting.

In a few minutes' time, you arrive at the floating boardwalk taking you into the heart of the Heron Pond Swamp. The green, duckweed-covered water is still; the soaring trunks of the bald cypresses stretch away into the distance as though they were a carefully planned demonstration of the laws of perspective. You may hear the staccato rhythm of a woodpecker in the distance. This place must not be lost.

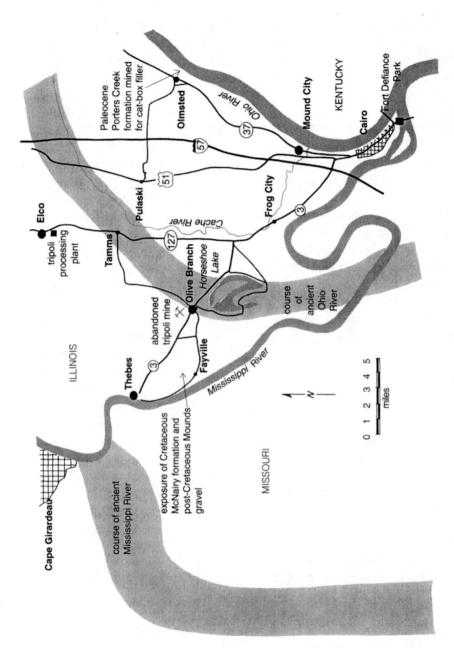

The southern tip of Illinois where the Ohio and Mississippi Rivers meet. The lightly shaded regions are the paths of the ancient rivers.

— 37 —
The Meeting of the Waters
OLMSTED TO CAIRO
Pulaski and Alexander Counties

To the River! To that portal of Eternity.
—Charles Dickens, *The Chimes*

There is more than a little irony in the use of this Dickens quotation. If any human being ever detested the city of Cairo and its environs more than this Victorian novelist, that person never wrote about it, at least not as scathingly. Yet the narrow wedge of land between the Mississippi and Ohio Rivers is an intrinsically fascinating locale, and not just to geology enthusiasts. Visitors from many parts of the world come to the waterfront park at Cairo's southern tip to pay homage to one of this planet's great meeting places.

The towns and countryside in the northern approach to Cairo are less frequently seen by interstate travelers, who hurry on to their next destination. But these places, too, have much to commend them to the student of earth history. To begin your circuit of the area, proceed to the Ohio River town of Olmsted, which is home to one of the state's most unique industries. If you make your way to the south side of town, you will find the quarries and processing plant of the Golden Cat Corporation, a company that markets a product that seems to have become a vital component of modern civilization—cat-box filler. While the quarries are off limits to the public, you can get a fairly good glimpse of them from your car window. The material being extracted here is not Pennsylvanian coal or Mississippian limestone; in fact, it isn't any sort of rock. It is a thick deposit of unconsolidated sediments, layers of white, tan, and dark gray clay sometimes referred to as fuller's earth. This clay, the primary constituent of the Porters Creek formation, dates from the Paleocene epoch—from the opening chapter of the Tertiary period, when the dinosaurs had just given up the stage to the rapidly evolving mammals. The climate of the earliest part of the Tertiary was much as it had been in the latter part of the preceding period, the Cretaceous. Illinois occupied an almost seasonless, subtropical setting. Excellent fossil remains found in other deposits in the area hint at luxuriant forest growth. One expert has likened this environment to present-day

An exposure of the Paleocene Porters Creek formation at an abandoned quarry in Olmsted.

Central America and Vietnam. Then, as now, the flowering plants dominated the scene—even though the grasses and other advanced types had not yet come into being. At this point, southern Illinois was situated at the end of an arm of the sea that extended northward along a troughlike depression, the Mississippi Embayment. The Porters Creek formation contains sharks' teeth, fish scales, and other fossils that prove it was laid down in a marine environment.

The embayment had begun to subside late in the Cretaceous period, and it continued to through the early Tertiary's Paleocene and Eocene epochs—as the several hundred feet of uncemented sediments in this region attest. Of the formations laid down in this interval from 90 million to 34 million years ago, only the Porters Creek clay has the optimum grease-absorbing properties deemed suitable for pet-box filler and for oil-spot removers used on the floors of auto repair shops. In keeping with the custom of nicknaming significant spans of geologic time—the Mesozoic era, for instance, was the Age of Reptiles—I suggest that the Paleocene be informally dubbed the Age of Kitty Litter. Were it not for what took place here during this one small slice of geologic time, cats of the present epoch would probably be much less pleased with the human beings who serve them.

272

Another unique mining industry centered in the southwestern tip of Illinois is devoted to the production of tripoli, a microcrystalline silica substance found in the Devonian bedrock that outcrops above the Mississippi Embayment deposits, some 15 miles to the west and northwest of Olmsted. Tripoli is a substance composed of tiny but very tough quartz particles that may have been left behind as residue when the calcite of silica-rich limestones was removed by groundwater. It is employed in a variety of industrial applications, including lens grinding and scouring compounds, and it has a more popular use as pool-cue chalk.

If you proceed from Olmsted south on Route 37 and then northwest on Route 3 to Olive Branch, and then go about a mile on the road heading northeast out of town, you will see an old tripoli-producing site on private property, where limestones of the Devonian Bailey formation (see essay 32) form the massive bluff of the abandoned valley of the ancient Ohio. The valley, now appropriated by the modest Cache River, is bounded on its northern side by a highland of Paleozoic rocks. On its southern side, the zone of yielding Cretaceous sediments forms no bluff at all. While the mine before you here is long defunct, there are active operations farther north, in the vicinity of the tripoli processing plant in Elco. These mines are worked

An old tripoli mining site near Olive Branch. The limestone exposed here belongs to the Devonian Bailey formation. This cliff is part of the high bluff that formed the northern limit of the ancient Ohio River valley.

in the most productive Devonian formation, the Clear Creek, which lies above the Bailey and Grassy Knob formations.

Another site in the Olive Branch area offers you an especially good look at the unconsolidated sediments that were deposited in this region in Cretaceous times and later. Take the Fayville Road due west from Olive Branch. At about 2.9 miles the pavement ends; continue straight on the gravel road. After an additional 0.5 mile the road bears right, and you'll soon see a large and deeply gullied hillside—powerful testimony to the force of erosion, where the earth's surface is mantled in yielding, uncemented sands and gravels, rather than in more resistant rock. Another 0.3 mile up the road, you will spot a small borrow pit on the right in which these loose sediments are well exposed. The white sand that forms most of the cut face belongs to the Cretaceous McNairy formation. Above it, at the very top, lies the reddish, chert-pebbled Mounds gravel. The Mounds is most definitely not a part of the Mississippi Embayment deposits that include the McNairy formation; in fact, it is much younger. While its exact age remains in doubt, largely because of its poor fossil content, it is clear that it was brought here by a precursor of the Mississippi River that flowed from the north. (Recent comparison to similar sediments farther south suggests that the Mounds may range in age from late in the Tertiary period's Miocene epoch to early in the Quaternary period's Pleistocene epoch.) Whatever its precise origin, it makes its presence obvious here by partly staining the pale sands below it with its own characteristically reddish hue. That ruddy color signals that it is rich in iron-bearing compounds.

Return by the same road to Olive Branch. Just south of town lies yet another type of geologic marvel: a large abandoned meander of the Mississippi River. This is Horseshoe Lake, home of peaceful bald-cypress groves and thousands upon thousands of not-so-peaceful, overwintering Canada geese supported by corn grown specifically for them in nearby fields. You can get a good look at this sportsman's paradise by heading down Route 3 and turning right at the entrance sign.

Water-filled, abandoned meanders—geologists call them oxbow lakes— are a common sight along the Mississippi Embayment section of the Mississippi. They form when this mature, winding river takes a permanent shortcut by cutting through the neck of one of its channel loops. This tendency of the Father of the Waters to shorten his own course once prompted Mark Twain to calculate that the river's lower half would eventually be only one and three-quarters miles long—with the result that Cairo and New Orleans would be able to share one mayor, one street system, and a single board of aldermen. When it becomes detached, the meander no longer is the conduit for the relatively fast-flowing water of the main channel. Instead, it is transformed into a curving, narrow, slackwater pond that supports a diverse community of plants and animals. Horseshoe Lake has that definite loop shape; but it is divided into two parts, more or less, by a central island. At its deepest, the bed of the lake lies only 6 feet beneath the languid surface.

Farther down Route 3, in the vicinity of the justly named Frog City, the terrain becomes downright swampy. This land has a long record of periodic submergence under the waters of the Mississippi—hence the extensive system of modern levees along the road. As you reach the outskirts of Cairo, you may feel as though you're entering the Cajun country of southern Louisiana. Indeed, you are now almost 40 percent of the way from Chicago to New Orleans. At one time, Cairo was the rising star of the Midwest. On paper, at least, it seemed to have the brightest prospects of any city in the world: it stood like an unavoidable tollgate at the confluence of the two greatest river transportation systems of a vigorous young nation. The surrounding countryside was brimming with agricultural potential and natural resources. Unfortunately, though, the euphoric hopes that Cairo would become the hub of the continent died long before the Civil War. It was all too apparent how prone to extensive flooding the

A barge carrying Illinois bituminous coal awaits its tug at Cairo's Fort Defiance Park, near the junction of the Mississippi and Ohio Rivers.

site was. Before its reputation for disaster spread, however, many people—
including speculators in Great Britain—forfeited their savings in unsavory
investment schemes.

Charles Dickens may have been one of the transatlantic investors who
lost money in Cairo's false dream. When he visited it in the 1840s, he was
aghast at what he saw:

> At the junction of the two rivers, on ground so flat and low and marshy,
> that at certain seasons of the year it is inundated to the house-tops, lies a
> breeding-place of fever, ague, and death; vaunted in England as a mine of
> Golden Hope, and speculated in, on the faith of monstrous representations,
> to many people's ruin. A dismal swamp; on which the half-built houses rot
> away . . . ; the hateful Mississippi, circling and eddying before it, and turning
> off upon its southern course a slimy monster hideous to behold; a hotbed
> of disease; an ugly sepulchre, a grave uncheered by any gleam of promise:
> a place without one single quality, in earth or air or water, to commend it:
> such is this dismal Cairo.

You may not share this poisonous opinion when you reach Cairo's
southern point, Fort Defiance Park. You can drive out to the Riverboat
Memorial on the point and survey the meeting of the two great rivers. The
state's official southern tip is an island west by southwest of here, but the
vantage point there couldn't be any better than this.

Before you stretches the great inland waterway that has shaped American settlement patterns and civilization for the past two centuries, just as
it has drained much of the continent's fresh water for thousands of years.
At this, our final site, there are no rock outcrops, no fault traces, no carefully
laid out displays of fossils or minerals. But this breezy place offers the most
important geology lesson of all. You can comprehend this lesson by watch-
ing the busy traffic of river barges almost at eye level, or by tracking the
antlike cars on the great bridges above you. Behind all this motion are the
ghosts of thousands of men, women, and children who have based their
migrations on where the water and valleys would take them. This, then, is
the lesson they have demonstrated: however much we shape the earth, it
shapes us more. Despite our propensity for change and our imagined in-
dependence, we remain the products of this planet and its processes. Our
culture and our individual lives are molded by forces and features of ancient
origin. And as these ceaselessly flowing waters reveal, we live not at the end
of geologic time, but in its midst. The earth moves on, as vitally as ever.

— Glossary —

Age of Reptiles. A popular nickname for the Mesozoic era.

aggrading stream. A stream that receives more sediment than it can remove. The channel of an aggrading stream is built up and not eroded downward.

alluvium. A general term for sediments laid down in stream valleys.

anticline. A structure in which rock beds have been arched or folded upward.

anticlinorium. A large-scale structure of the earth's crust, composed of at least several anticlines.

aquifer. A permeable zone of rock or unconsolidated sediments that acts as a ground-water reservoir.

backdune zone. As used in this book, the area of older dunes that lie well inland of a shoreline's modern beach ridge.

barchan dune. A crescent-shaped dune with its "horns" pointing downwind.

basalt. A fine-grained and usually dark-toned igneous rock.

base level. The lowest possible level to which streams in a particular area may cut downward and erode the landscape around them.

beach ridge. An elongated and often continuous dune ridge that forms just behind a shoreline's beach and surf zone.

bedded deposit. As used in this book, an essentially horizontal fluorspar deposit that lies parallel to the bedding of the surrounding rock units.

Bedford limestone. The trade name for a Mississippian-period, biocalcarenite limestone quarried in southern Indiana and much used as a building stone in Chicago and elsewhere.

biotite. A dark-colored silicate mineral often found in igneous rocks.

bitumen. An often dark-colored and tarlike substance composed of hydrocarbon molecules derived from the chemical breakdown of organisms.

bog. This term is often used to denote any inaccessible or unexploitable wetland. Life and earth scientists have a more precise definition: a bog is a nutrient-poor peatland characterized by highly acidic conditions. Its surface stands above water level due to the gradual buildup of the peat. Bogs are inhabited by plants (including certain sphagnum-moss species and leatherleaf shrubs) that are specially adapted to these challenging conditions.

breccia. A sedimentary rock composed of coarse, angular fragments in a finer matrix.

calcareous. Having a high calcium carbonate content. Generally, calcareous rocks are defined as those that are at least 50 percent calcium carbonate.

calcite. The mineral form of calcium carbonate.

Canadian Shield. A large area in Canada and parts of Wisconsin, Minnesota, and Michigan's Upper Peninsula, where Precambrian (and therefore very ancient) rocks are exposed at the earth's surface.

carbonate rock. A sedimentary rock type mainly composed of a carbonate mineral. The primary examples are limestone (calcium carbonate) and dolomite (magnesium carbonate).

catastrophism. A term with various definitions and ramifications, but which often refers to a geologic doctrine in which the evolution of the earth is taken to be the result of sudden, large-scale, dramatic events rather than of subtler, more gradual processes.

channel lag. Gravel or other coarse sediment that moves along a streambed more slowly than the finer, suspended material borne along in the current.

chert. A hard sedimentary rock composed of quartz and other related silicate minerals. For all intents and purposes, "flint" is synonymous.

cladding. In architecture, a material (such as ornamental stone) used to cover and decorate a building's exterior.

clay. A sediment type composed of particles less than 1/256 millimeter in diameter.

concretion. A lozenge-shaped or globular stone nodule that forms around a nucleus. Concretions found in Pennsylvanian-period strata often contain exquisitely rendered plant or animal fossils; they are much sought after by collectors.

continental drift. Sometimes used as a general term denoting the movement of continents over geologic time. Geologists tend to restrict the term to the early-twentieth-century theory of Alfred Wegener, which has been largely modified and overtaken by the theory of plate tectonics.

continental ice sheet. A glacier that forms and spreads over large areas of often low-lying continental terrain—in contrast with an alpine glacier, which is formed in a mountainous region and which is mostly confined to mountain valleys.

crinoid. A bottom-dwelling, invertebrate marine animal that has a main stem, root-like structures, and other plantlike features—hence, its common nickname of "sea lily."

crossbedding. A distinctive pattern found in some sedimentary rocks. The beds—sometimes very narrow—lie at angles that are different from the main bedding plane.

crust. The uppermost section of the earth. It lies above the mantle and contains the continents and ocean floors.

cyclothem. An ideal repeating pattern of Pennsylvanian-period strata that indicates a cyclical return of such various conditions as seawater retreat, coal-swamp, and seawater invasion.

delta. The often wedge- or fan-shaped accumulation of sediments deposited at the mouth of a stream.

diabase. A dark-toned igneous rock.

disconformity. A boundary between different, parallel rock beds that represents a gap in geologic history, due to an intervening phase of erosion or at least no deposition.

dolomite. A term with two meanings: the carbonate mineral that contains both calcium and magnesium; and the sedimentary rock that contains at least 50 percent of the mineral dolomite.

dolostone. A term used by some geologists to denote the rock type more generally known as dolomite—to distinguish it from the mineral of the same name. The term is not in general use in Illinois.

drift. *See* **glacial drift.**

Driftless Section. *See* **Wisconsin Driftless Section.**

dune. A pile of sand deposited by the wind. It may take the form of a hill, a mound, or a ridge.

280

end moraine. A ridge composed of till that was created by a glacier that was melting at its margin as quickly as it was replenished with ice from its rear.

epoch. With regard to the geologic time scale, the major subdivision of a period.

era. A major subdivision of the geologic time scale. Each era is composed of at least two periods; the Paleozoic era contains seven periods.

erratic. A rock detached and carried from its parent outcrop by a glacier.

esker. A winding ridge composed of stratified gravel and sand. An esker forms when waterborne sediments fill a tunnel or other channel beneath, in, or on top of a stationary ice sheet.

fault. A breach in the earth's crust where rock units have been displaced along a plane. On a map, the surface expression of a fault is a straight or curving line.

fault zone. A weakened area of the earth's crust that contains more than one fault.

feldspar. Any of a group of silicate minerals. Feldspars are often found in igneous rocks; it has been estimated that they make up more than half the earth's crust.

floodplain. The flat surface of a valley bottom composed of sediments deposited by the valley's stream. When the stream is confined to its channel, the floodplain lies a little above the water surface; during floods, it is inundated unless it is protected by levees.

fluorspar. The common and trade name for the mineral fluorite (calcium fluoride). Fluorspar is often a handsome tone of purple or blue, but it can also be other colors, or clear.

foredune zone. As used in this book, the area of young dunes that lie close to a shoreline's modern beach ridge.

formation. An assemblage of strata that is the primary unit of stratigraphic classification.

geode. A more or less spherical body with a hollow center and an interior surface covered with crystals.

geomorphology. The earth science of landforms and the processes that create and act upon them.

geophysics. The earth science that uses the quantitative concepts and techniques of physics to determine underground structural features and the forces they generate.

glacial drift. An old but still widely used term for any sediment or rock debris transported and dumped by an ice sheet or its meltwaters.

glacial striation. A groove or scratch on a rock surface etched by another rock borne by a moving ice sheet.

glaciation. An episode in earth history consisting of the creation, advance, and retreat of an ice sheet. The term is not synonymous with "ice age," which is taken to be a larger episode containing more than one glaciation.

graben. A troughlike depression in the earth's crust that is bounded on both sides by faults.

granite. An igneous rock, often whitish or pinkish in overall tone, made up of readily visible crystals.

greenstone. A green-toned metamorphic rock.

groin. A barrier placed perpendicular to a shore, to inhibit beach erosion.

ground moraine. A low, till-mantled surface created by a rapidly retreating glacier.

groundwater. Water contained under the earth's surface in rock strata or sediments.

group. A major unit of stratigraphic classification composed of more than one formation.

gypsum. A transparent to opaque sulfate mineral that often forms from the evaporation of a body of water.

hanging wall. The rock units bordering a fault that lie over the fault itself.

hematite. A mineral form of iron oxide. It ranges in color from a reddish tone to a metallic silver.

hill prairie. A fascinating and beautiful habitat found on some loess-blanketed Illinois river blufftops.

hornblende. A dark-colored silicate mineral often found in igneous rocks.

ice age. A general term for any major episode in earth history consisting of a series of glaciations. When capitalized, "the Ice Age" is used as a popular name for the Pleistocene epoch—which is only the latest of at least several ice ages that have occurred over geologic time.

igneous rock. Any rock that formed from the cooling of magma.

Illinoian stage, Illinoian glaciation. The Pleistocene-epoch glaciation that occurred approximately from 300,000 to 125,000 years ago.

Illinois Basin. Illinois's largest geologic structural feature, centered in the southeastern section of the state. It is a huge, oval, down-arched accumulation of strata formed over many millions of years, in the Paleozoic era.

interglacial. Used as a noun, the stage of overall warmer conditions between glaciations.

ironstone concretion. *See* **concretion.**

joint. A fracture or gap in a rock, in which no significant displacement of the rock has occurred.

Joliet marble. An old trade name for the Sugar Run dolomite quarried in Joliet and once used widely as building stone in Chicagoland. In fact, the stone is not marble because it is not a metamorphic form of carbonate rock.

kame. A conical or ridgelike hill, most often composed of stratified sand or gravel, that is formed in one of several ways: as a delta at the front of a melting glacier, or as an accumulation of sediments either in a depression in the ice or on the flanks of a glacier confined to a river valley.

Kansan stage, Kansan glaciation. An obsolete term for what was once thought to be the second of the Pleistocene epoch's four glaciations. It is now known that there were many glaciations—perhaps up to twenty. The Kansan is now included in the Pre-Illinoian stage.

karst topography. A limestone or dolomite terrain that has well-developed and widespread sinkholes, caverns, and subsurface streams.

kettle. A depression created when a detached block of glacier ice, at least partially embedded in surrounding glacial drifts, finally melts away. Kettles often contain ponds, peatlands, or marshes.

klint, *plural* **klintar.** A resistant hill or mound composed of an ancient fossil reef.

lacustrine. Referring to a lake or to a feature produced by a lake.

landform. Any feature on the earth's surface that is produced by natural forces: for example, a mountain, a lake, a moraine, a dune, a ravine.

lime. Calcium oxide or a closely related chemical compound.

limestone. A sedimentary rock composed of at least 50 percent calcite.

limonite. An iron-bearing oxide mineral that is often yellowish.

lithographic limestone. A very fine-grained limestone that was once used for engraving plates. Also known as "micrite."

loess. Pronounced *luss*. A tan, windblown silt that is often found in thick deposits on Mississippi and Illinois River bluffs, as well as on other terrains.

longitudinal dune. A long, ridgelike dune oriented parallel to the prevailing wind direction.

longshore current. A current generated by waves that run into the shoreline at an angle. Along the Illinois coast of Lake Michigan, the waves most often hit the shore moving southwestward; consequently, the prevailing longshore current runs down the coast, southward.

magma. Molten rock.

mantle. The large area of the earth's interior that lies between the crust and the core.

marble. Metamorphic carbonate rock that consists mainly of recrystallized calcite or dolomite.

marsh. A type of wetland characterized by an absence of peat and by a dominant plant community composed of herbaceous plants rather than of trees and shrubs.

mature stream. A stream that has reached the stage in which downcutting is essentially replaced by sidecutting, and in which the stream has reached an equilibrium where it carries away only as much sediment as it receives from upstream.

member. A unit of stratigraphic classification that is the primary subdivision of a formation.

metamorphic rock. Rock formed when sedimentary or igneous rock is altered but not completely melted by increased heat or pressure.

Mississippi Embayment. The huge, down-arched accumulation of sediments located in the lower Mississippi River valley, from southernmost Illinois to the Gulf of Mexico. Also known as the Embayment syncline.

moraine. In general, this term is synonymous with **end moraine.** *See also* **ground moraine.**

Nebraskan stage, Nebraskan glaciation. An obsolete term for what was once thought to be the first of the Pleistocene epoch's four glaciations. It is now known that there were many glaciations—perhaps up to twenty. The Nebraskan is now included in the Pre-Illinoian stage.

old stream. A stream that has virtually reached base level and which has developed elaborate meanders across its floodplain.

oolite. A limestone composed of tiny, cemented spherical grains that each formed around a sand particle or other type of nucleus.

outcrop. A rock exposure.

outwash. Sand or gravel deposited at the edge of a melting glacier or in front of an end moraine.

outwash plain. A wide, gently sloping area of outwash deposited in front of a melting glacier.

overburden. An overlying zone of rock or unconsolidated sediment that must be removed to reach rock to be mined.

paleobotanist. A scientist who studies fossil plants, algae, and fungi, as well as their evolutionary significance.

paleontologist. A scientist who studies fossil organisms—especially fossil animals—as well as their evolutionary significance.

Pangaea. The ancient supercontinent that existed in late Paleozoic and early Mesozoic time. Plate-tectonics theorists have suggested that Pangaea stretched from one polar region to the other.

parabolic dune. A crescent-shaped dune that has its "horns" pointing upwind.

peat. The only partially decomposed organic material that is found in some wetland environments.

peatland. A wetland in which the soil is mainly peat.

peneplain. A landscape that has been eroded down to a low, rolling surface close to base level.

period. Regarding the geologic time scale, the major subdivision of an era.

phreatic. Pertaining to caverns in karst terranes that form in the zone of groundwater saturation under the water table.

physiographic province. A region characterized by similar climate, landforms, and subsurface structures.

physiography. In modern earth-science usage, the study of landforms. Now usually considered synonymous with **geomorphology.**

plate tectonics. The theory that addresses the now-abundant evidence of continental movement and seafloor spreading by describing the earth's crust as a series of "plates" that interact with each other, and also with the underlying mantle.

Precambrian basement. The complex of Precambrian igneous rocks that underlies Paleozoic sedimentary rocks, often at great depth, throughout Illinois. No Precambrian rocks outcrop in the state.

Pre-Illinoian stage, Pre-Illinoian glaciations. A blanket term for all the Pleistocene-epoch glaciations before the Illinoian stage. The Pre-Illinoian stage lasted approximately from 1.6 million to 300,000 years ago.

quartz. A hard and common rock-forming mineral that is a crystalline form of silica (silicon dioxide).

quartzite. An extremely hard rock that is a metamorphosed sandstone.

radiocarbon dating. A technique for estimating the age of ancient deposits or features by measuring the amount of the radioactive isotope carbon-14 found in organic material. This technique has an effective dating range from approximately 500 years ago to a maximum of about 40,000 years ago.

ravine. A small, narrow, steep, and usually wooded stream valley, which is V-shaped in cross section.

roadcut. An outcrop exposed by roadway construction.

rock shelter. An open-mouthed cavity at the base of a rocky cliff, such as those formed in Pennsylvanian sandstone at several sites in the unglaciated portion of southern Illinois.

rubble bar. A bar in a river or flood channel composed of cobbles and rock rubble rather than finer sediments such as sand.

sandstone. A sedimentary rock composed of cemented sand grains.

Sangamonian stage, Sangamonian interglacial. The interglacial between the Illinoian and Wisconsinan glaciations. The Sangamonian stage lasted approximately from 125,000 to 75,000 years ago.

sea lily. *See* **crinoid.**

sedimentary rock. Rock formed from the lithification of deposited sediments (as in sandstone or shale) or from chemical precipitation (as in some limestones).

serpentine. A striking dark green rock with wavy light veins. It is formed as part of an ophiolite sequence and, according to a current theory, may be formed when mantle material at great depth in an oceanic spreading center comes into contact with seawater.

shale. Sedimentary rock formed from the lithification of clay particles.

shatter cone. A fractured, cone-shaped rock fragment found at meteorite impact sites.

silica. Silicon dioxide. A common mineral form of silica is quartz.

silicate mineral. Any of the large group of minerals containing the chemical element silicon.

silt. A sediment particle no less than 1/256 of a millimeter and no greater than 1/16 of a millimeter.

siltstone. Sedimentary rock composed of silt particles.

sinkhole. A round or oval depression in the surface of a limestone or dolomite terrane caused by the slow dissolving of carbonate rock or by subsurface collapse.

spit. An elongated and usually curved body of sand or gravel built in shoreline water.

spreading center. A more or less linear, deep-ocean zone that is the border of two plates. Magma erupts along this zone to create new seafloor and to slowly widen the ocean basin.

stratified. Occurring in bands, beds, or layers.

stratigraphy. The earth science that deals with the description, chronology, and interrelationships of rock strata.

stratum, *plural* **strata.** A layer or bed of sedimentary rock.

subglacial channel. A drainageway or stream course formed under an ice sheet.

supercontinent. A very large landmass formed from the docking of several continents.

syncline. A structure in which rock layers have been arched or folded downward.

terrace. A horizontal, bench- or platformlike surface. Terraces are the remnants of old floodplains.

terrain. The combination of landforms making up a region's surface.

terrane. A large-scale, three-dimensional assemblage of rock sections that form both a region's surface and its subsurface.

till. An unstratified jumble of rock debris deposited directly by an ice sheet (not by meltwater issuing from the ice sheet). Till sediments range in size from clay particles to boulders.

till plain. A till-blanketed plain of very low relief.

travertine. A sedimentary rock that ranges from white to tan and is often marked with bands or series of small cavities. Travertine forms from the precipitation of calcium carbonate in hot springs and zones of seeping groundwater.

unconformity. A boundary between two rock units of different ages, representing a gap in the geologic record.

underclay. The band of clay directly under a coal bed. The underclay represents the soil where ancient plants that ultimately became the coal once grew.

underfit stream. A stream too small to have formed the valley it occupies.

uniformitarianism. While it can be defined in more than one way and with varying emphasis, it is generally taken to be the geologic doctrine holding that the earth has always been shaped by forces that can be witnessed today. Often linked with this is the belief that catastrophes have played a relatively minor role in earth history, while gradual, undramatic change has dominated the geologic record.

upsection. The direction from older to younger strata. In undisturbed, horizontal beds, this direction is up.

vadose. Pertaining to caverns in karst terranes that form above the water table.

valley train. Outwash from a moraine or melting glacier that has been deposited in a stream valley.

vein deposit. As used in this book, fluorspar deposits that formed in vertical or near-vertical fault fissures.

water table. The upper boundary of a zone saturated with groundwater.

wetland. An area where water is at or near the surface throughout the year.

Wisconsin arch. A major structural feature in Wisconsin, in which strata have been arched upward along a lengthy north-south axis.

Wisconsin Driftless Section. A hilly region of southwestern Wisconsin and north-western Illinois that was apparently bypassed by the Pleistocene ice sheets.

Wisconsinan stage, Wisconsinan glaciation. The final glaciation of the Pleistocene epoch. It lasted approximately from 75,000 to 10,000 years ago.

xenolith. A foreign body of rock contained in a surrounding igneous rock.

youthful stream. A stream characterized by active downcutting and by a steep valley that is V-shaped in cross section.

— Annotated Bibliography —

Besides using the books, papers, and maps listed below, you can find more information about the geology of Illinois in Illinois State Geological Survey (ISGS) Field Trip Guidebooks. United States Geological Survey (USGS) 7.5-minute topographic quadrangle maps show locations of the sites described in this text. Selected titles in the ISGS Field Trip series, together with many USGS topographic maps, can be purchased at the Public Information Office of the Illinois State Geological Survey, 615 East Peabody Drive, Champaign, 61820 (on the University of Illinois campus).

Quaternary and Glacial Geology

Chrzastowski, Michael J., Todd A. Thompson, and C. Brian Trask. 1994. Coastal geomorphology and coastal cell divisions along the Illinois-Indiana coast of Lake Michigan. *Journal of Great Lakes Research* 20:27–43.

Colman, S. M., et. al. 1994. Deglaciation, lake levels, and meltwater discharge in the Lake Michigan basin. *Quaternary Science Review* 13:879–90.

Dreimanis, Aleksis. 1977. Late Wisconsin glacial retreat in the Great Lakes region, North America. *Annals of the New York Academy of Sciences* 288:70–89.

Erickson, Jon. 1990. *Ice Ages: Past and Future.* Blue Ridge Summit, Pa.: TAB Books.

Hajic, Edwin Robert. 1990. *Late Pleistocene and Holocene Landscape Evolution, Depositional Subsystems, and Stratigraphy in the Lower Illinois River Valley and Adjacent Central Mississippi River Valley.* Doctoral dissertation, University of Illinois, Urbana-Champaign, Illinois.

Hansel, Ardith K., and W. Hilton Johnson. 1992. Fluctuations of the Lake Michigan lobe during the late Wisconsin subepisode. *Geologiska Undersökning* 81:133–44. Available as Illinois State Geological Survey Reprint 1993F.

Hansel, Ardith K., et. al. 1985. Late Wisconsinan and Holocene history of the Lake Michigan basin. Geological Society of Canada Special Paper 30, pp. 39–53. An excellent and up-to-date chronology of Wisconsinan glacial retreat, ancient Lake Chicago, and the creation of modern Lake Michigan. A must for the northeastern-Illinois naturalist. This paper was recently reprinted as an appendix in Palmer's *Wisconsin's Door County: A Natural History* (see below)—reason enough for an Illinois resident to acquire that book.

Horberg, Leland. 1918. *Preglacial Drainage Systems in Illinois.* Map. 1948 Reprint. Urbana: Illinois State Geological Survey.

Imbrie, John, and Katherine Palmer Imbrie. 1979. *Ice Ages.* Cambridge: Harvard University Press. An excellent, nontechnical account of the development of theories about ice ages. The book makes a good complement to Sharp's *Living Ice,* which is more of a straight description of glacial processes.

Johnson, W. Hilton, David W. Moore, and E. Donald McKay III. 1986. Provenance of late Wisconsinan (Woodfordian) till and origin of the Decatur sublobe, east-central Illinois. *Geological Society of America Bulletin* 97:1098–1105. Available as Illinois State Geological Survey Reprint 1987A.

Killey, Myrna M. 1974. Erratics are erratic. Geogram 2. Urbana: Illinois State Geological Survey. An excellent briefing sheet on the glacial erratics found in many places in Illinois.

Lineback, Jerry A. 1979. *Quaternary Deposits of Illinois.* Map. Urbana: Illinois State Geological Survey. A beautiful and informative wall map.

Lineback, Jerry A., et al. 1979. *Wisconsinan, Sangamonian, and Illinoian Stratigraphy in Central Illinois.* Illinois State Geological Survey Guidebook 13. Urbana: Illinois State Geological Survey. A technical text for specialists, with detailed outcrop descriptions for selected sites from the Peoria area to Springfield.

Martin, Paul S., and Richard G. Klein, eds. 1984. *Quaternary Extinctions.* Tucson: University of Arizona Press. A voluminous and scholarly study of the great disappearance of many mammalian species during the Ice Age.

Ogden, J. Gordon III. 1977. The late Quaternary paleoenvironmental record of northeastern North America. *Annals of the New York Academy of Sciences* 288:16–34.

Pielou, E. C. 1991. *After the Ice Age: The Return of Life to Glaciated North America.* Chicago: University of Chicago Press. A Canadian naturalist explains how plants, animals, and ecosystems have coped (or have not coped) with the climatic upheavals of the Quaternary period.

Schneider, Allan F., and Ardith K. Hansel. 1990. Evidence for post–Two Creeks age of the type Calumet shoreline of glacial Lake Chicago. Geological Society of America Special Paper 251.

Sharp, Robert P. 1988. *Living Ice: Understanding Glaciers and Glaciation.* New York: Cambridge University Press. Engagingly written and well illustrated, this nontechnical guide is one of the best introductions to glacial processes and landforms.

Willman, H. B., and John C. Frye. 1970. *Pleistocene Stratigraphy of Illinois.* Illinois State Geological Survey Bulletin 94. Urbana: Illinois State Geological Survey.

———. 1970. *Woodfordian Moraines of Illinois.* Map. Urbana: Illinois State Geological Survey. This map is still considered accurate, except for the "Snoopy" ice-limit bulge shown in the Oregon-Dixon area and for the exact extent of the Valparaiso morainic system. The Wisconsinan moraine names used in this book are largely taken from this map.

Paleontology, Paleobotany, and Evolution Issues

Andrews, Henry N. 1980. *The Fossil Hunters: In Search of Ancient Plants.* Ithaca, NY: Cornell Univ. Press. An eminent paleobotanist traces the history of his science through the lives of its leading proponents. A great deal of exciting research has been done since this book was written, and a revised version or sequel is badly needed.

Beck, Charles B., ed. 1976. *Origin and Early Evolution of Angiosperms.* New York: Columbia University Press.

Behrensmeyer, Anna K., et al. 1992. *Terrestrial Ecosystems Through Time.* Chicago: University of Chicago Press. This symposium spin-off has a particularly good section on the ecology of Pennsylvanian coal swamps.

Colbert, Edwin H. 1973. *Wandering Lands and Animals.* New York: E. P. Dutton. A popular account of how continental drift affected the distribution of fossil animals.

Eiseley, Loren. 1961. *Darwin's Century.* Garden City, N.Y.: Anchor Books.

Eldredge, Niles. 1995. *Dominion.* New York: Henry Holt. A discussion of humankind's Pleistocene origins, and of its current overpopulation dilemma.

Gould, Stephen Jay. 1989. *Wonderful Life: The Burgess Shale and the Nature of History.* New York: W. W. Norton. A thought-provoking look at the vicissitudes of evolution, using one of the world's most fascinating assemblages of fossils as its main theme. As with practically all of Gould's books, it is lucidly written and justly appealing to the curious layperson.

Jennings, James R. 1990. *Guide to Pennsylvanian Fossil Plants of Illinois*. Educational Series 13. Champaign: Illinois State Geological Survey. An informative booklet with good illustrations of representative fossil-plant types. It gives excellent fossil-hunting and collecting tips. These include a list of locales where the fossils are most plentiful.

Lowenstam, Heinz. 1948. *Biostratigraphic Studies of the Niagaran Inter-reef Formations in Northeastern Illinois*. Scientific Papers Vol. 4. Springfield: Illinois State Museum.

Matten, Lawrence C. 1971. Tracking down the petrified remains of the Illinois of 500 million years ago. *Outdoor Illinois* January:19–26. As printed, the title of this fascinating article apparently contains a typographic error: *50* million years makes more sense. A good rundown on Mississippian, Cretaceous, and Tertiary fossil-plant finds in southern Illinois.

Moore, Raymond C., Cecil G. Lalicker, and Alfred G. Fischer. 1952. *Invertebrate Fossils*. New York: McGraw-Hill.

Page, David. 1861. *The Past and Present Life of the Globe*. Edinburgh, Scotland: Wm. Blackwood and Sons. A delightful glimpse at Victorian paleontology and geology, replete with a defense of divinely directed evolution and small but flamboyant illustrations, each like the nightmares of children. Now long out of print.

Phillips, Tom L., Hermann W. Pfefferkorn, and Russell A. Peppers. 1973. *Development of Paleobotany in the Illinois Basin*. Illinois State Geological Survey Circular 480. Urbana: Illinois State Geological Survey. For the student of fossil plants, this historical account is a must, since it describes the development of a science in one of the world's most important paleobotanical regions.

Raup, David M. 1991. *Extinction: Bad Genes or Bad Luck?* New York: W. W. Norton. A thoughtful yet entertaining analysis of modern extinction theories. Not technical.

Stanley, Steven M. 1987. *Extinction*. New York: Scientific American.

Stewart, Wilson N., and Gar W. Rothwell. 1992. *Paleobotany and the Evolution of Plants*. 2nd ed. New York: Cambridge University Press.

Taylor, Thomas N., and Edith L. Taylor. 1993. *The Biology and Evolution of Fossil Plants*. Englewood Cliffs, N.J.: Prentice-Hall.

White, Mary E. 1990. *The Flowering of Gondwana*. Princeton: Princeton University Press.

General Geology Texts and References

Adams, George F., and Jerome Wyckoff. 1971. *Landforms*. New York: Golden Books. A delightful little pocket guide, but unfortunately now out of print.

Agar, William M., Richard Foster Flint, and Chester R. Longwell, eds. 1929. *Geology from Original Sources*. New York: Henry Holt.

Bates, Robert L., and Julia A. Jackson, eds. 1984. *Dictionary of Geological Terms*. 3rd ed. New York: Anchor Books. One of the best geo-dictionaries in print today.

Billings, Marland P. 1972. *Structural Geology*. Englewood Cliffs, N.J.: Prentice-Hall.

Bretz, J Harlan. 1955. *Geology of the Chicago Region*. Parts I and II. Urbana: Illinois State Geological Survey. Out of print, but still available in some Chicagoland libraries. Understandably out of date in a few specifics, it nevertheless is an excellent survey of the geology of Illinois's most populous region.

Chronic, Halka. 1987. *Roadside Geology of New Mexico*. Missoula, Mont.: Mountain Press.

Compton, Robert R. 1962. *Manual of Field Geology*. New York: John Wiley and Sons.

Davis, William Morris. 1909. *Geographical Essays*. New York: Dover. Lucidly written by one of the founders of geomorphology. Still of great interest to professionals and interested laypersons.

Dobrin, Milton B. 1988. *Introduction to Geophysical Prospecting.* 4th ed. New York: McGraw-Hill.

Dolphin, R. E. n. d. The geology and landforms of Beall Woods. An unpublished paper submitted to Illinois Department of Conservation officials at Beall Woods State Park.

Dott, Robert H., Jr., and Roger L. Batten. 1993. *Evolution of the Earth.* 4th ed. New York: McGraw-Hill. A recommended undergraduate-level text on the earth's geologic history.

Dorr, John A., Jr., and Donald F. Eschman. 1970. *Geology of Michigan.* Ann Arbor: University of Michigan Press. This excellent, well-illustrated hardcover textbook is a bit out of date now, but it still makes a good introduction to the geology of both the Upper and Lower Peninsulas.

Easterbrook, Donald J. 1993. *Surface Processes and Landforms.* New York: Macmillan. A modern and highly recommended undergraduate-level geomorphology text.

Geologic Highway Map: Great Lakes Region. 1978. Tulsa: American Association of Petroleum Geologists. A generalized and rather small-scale bedrock map that includes Illinois is combined with text, detail maps, and other fascinating information. A geological treasure chest. To order this and other maps in the series, contact the AAPG at P.O. Box 979, Tulsa, OK 74101.

Geologic Highway Map: Mid-Atlantic Region. 1987. Tulsa: American Association of Petroleum Geologists.

Geologic Highway Map: Mid-Continent Region. 1986. Tulsa: American Association of Petroleum Geologists.

Geologic Highway Map: Northern Great Plains Region. 1984. Tulsa: American Association of Petroleum Geologists.

Goethe, Johann Wolfgang von. 1988. *Scientific Studies.* Vol. 12 of *The Collected Works.* Douglas Miller, trans. Princeton: Princeton University Press.

Gould, Stephen Jay. 1987. *Time's Arrow, Time's Cycle.* Cambridge: Harvard University Press. First-rate historical analysis of the worldviews of James Hutton and Charles Lyell.

Gray, Henry H., Curtis H. Ault, and Stanley J. Keller. 1987. *Bedrock Geology Map of Indiana.* Bloomington: Indiana Geological Survey.

Grieve, Richard, et. al. 1995. The record of terrestrial impact cratering. *GSA Today* 5:189–96.

Harris, Stanley E., Jr., C. William Horrell, and Daniel Irwin. 1977. *Exploring the Land and Rocks of Southern Illinois.* Carbondale: Southern Illinois University Press. A commendable and recommended work on sites of special geomorphological interest in the state's southern counties.

Hawker, Jon L. 1992. *Missouri Landscapes: A Tour Through Time.* Rolla, Mo.: Missouri Dept. of Natural Resources. A nontechnical guide to the geologic and biologic history of one of Illinois's neighbor states.

Hunt, Charles B. 1974. *Natural Regions of the United States.* San Francisco: W. H. Freeman.

Hurlbut, Cornelius S. 1971. *Dana's Manual of Mineralogy.* 18th ed. New York: John Wiley and Sons.

Jackson, Kern C. 1970. *Textbook of Lithology.* New York: McGraw-Hill.

Krumbein, W. C., and L. L. Sloss. 1963. *Stratigraphy and Sedimentation.* 2nd ed. San Francisco: W. H. Freeman.

LaBerge, Gene L. 1994. *Geology of the Lake Superior Region.* Phoenix: Geoscience Press. For an introduction to the Precambrian history of the upper Midwest, read this book. It also describes igneous and metamorphic rock types that can be found in Illinois as glacial erratics.

Lange, Kenneth L. 1989. *Ancient Rocks and Vanished Glaciers: A Natural History of Devil's Lake State Park, Wisconsin.* Stevens Point, Wisc.: Worzalla Publishing.

Laudan, Rachel. 1987. *From Mineralogy to Geology*. Chicago: University of Chicago Press. An excellent reassessment of the European development of the geological sciences before the time of Hutton, Playfair, and Lyell. Highly recommended for serious students and professionals.

Longwell, Chester R., Richard Foster Flint, and John E. Sanders. 1969. *Physical Geology*. New York: John Wiley and Sons.

Lyell, Charles. 1990. *Principles of Geology*. 3 vols. Facsimile of 1st ed. of 1830–1833. Chicago: University of Chicago Press. One of the most influential works in the history of science. Trained as a barrister, Lyell advocated Hutton's uniformitarianism with great skill and lucid prose.

Martin, Lawrence. 1932. *The Physical Geography of Wisconsin*. 2nd ed. Madison: Wisconsin State Geological and Natural History Survey.

Mather, Kirtley F., and Shirley L. Mason, eds. 1970. *A Source Book in Geology, 1400–1900*. Cambridge: Harvard University Press.

Mather, Kirtley, ed. 1967. *Source Book in Geology, 1900–1950*. Cambridge: Harvard University Press.

McPhee, John. 1989. *The Control of Nature*. New York: Farrar, Straus, and Giroux. This excellent collection of essays includes an account of human attempts to influence natural processes of the lower Mississippi River.

———. 1993. *Assembling California*. New York: Farrar, Straus, and Giroux. This latest book in McPhee's superb series on modern geology includes a description of the exotic origin of serpentine. Science writing at its best: informative, stylistically beautiful, and nontechnical.

Mikulic, Donald G. 1987. The Silurian reef at Thornton, Illinois. *Geological Society of America Centennial Field Guide, North Central Section*.

Mikulic, Donald G., and Joanne Kluessendorf. 1987. Ordovician-Silurian unconformity at Kankakee River State Park, Illinois. *Geological Society of America Centennial Field Guide, North Central Section*.

———. 1991. Illinois' State Fossil, *Tullimonstrum gregarium*. Geogram 10. Champaign: Illinois State Geological Survey.

———. 1994. The classic Silurian reefs of the Chicago area. *Kalamazoo 1994: Field Trips Guidebook*, pp. 194–244. Kalamazoo: Western Michigan University.

Miller, James Andrew. 1973. *Quaternary History of the Sangamon River Drainage System, Illinois*. Reports of Investigations No. 27. Springfield: Illinois State Museum.

Nelson, W. John. 1987. Horseshoe Quarry, Shawneetown fault zone, Illinois. *Geological Society of America Centennial Field Guide, North Central Section*, pp. 241–44.

Nelson, W. John, and Donald K. Lumm. 1990. *Geologic Map of the Stonefort Quadrangle, Illinois*. Champaign: Illinois State Geological Survey. It includes the Bell Smith Springs site in Shawnee National Forest.

Palmer, John C., ed. 1990. *Wisconsin's Door Peninsula: A Natural History*. Appleton, Wisc.: Perin Press.

Park, Charles F., Jr., and Roy A. MacDiarmid. 1970. *Ore Deposits*. 2nd ed. San Francisco: W. H. Freeman.

Pellant, Chris. 1992. *Rocks and Minerals*. Eyewitness Handbooks. New York: Dorling Kindersley. An excellent, beautifully designed picture guide with brief but pithy text sections.

Pennick, James Lal, Jr. 1981. *The New Madrid Earthquakes*. Rev. ed. Columbia: University of Missouri Press. A historian's account of the greatest North American earthquake episode on record.

Petersen, Morris S., J. Keith Rigby, and Lehi F. Hintze. 1980. *Historical Geology of North America*. 2nd ed. Dubuque, Ia.: William C. Brown. A relatively brief description of the earth's geologic past, especially well suited to the nonspecialist. Recommended.

Pough, Frederick H. 1976. *A Field Guide to Rocks and Minerals*. Peterson Field Guide Series. Boston: Houghton Mifflin.

Prothero, Donald R. 1994. *The Eocene-Oligocene Transition: Paradise Lost*. New York: Columbia University Press. A fairly nontechnical account of the worldwide cooling in the current, Cenozoic era. It includes important new research findings. An engaging look at an unjustly ignored but important part of the earth's past.

Reineck, Hans-Erich, and Indra Bir Singh. 1980. *Depositional Sedimentary Environments*. 2nd ed. Berlin: Springer-Verlag. This is a leading sedimentology text, but it is written (or at least translated) in relentlessly dry, awkward, technical prose. Also, its highly reflective, glossy paper strains the reader's eye.

Ritter, Dale F., R. Craig Kochel, and Jerry R. Miller. 1995. *Process Geomorphology*. 3rd ed. Dubuque, Ia.: William C. Brown. One of the leading modern geomorphology texts.

Schuberth, Christopher J. 1986. *A View of the Past: An Introduction to Illinois Geology*. Springfield: Illinois State Museum. Geologist and museum educator Schuberth deserves hearty praise for this short but solid overview of the state's geology. Written in terms the general public understands, this book proved to be extremely popular, even though it was one of the Illinois State Museum's few modern forays into publishing vibrant, nontechnical, public-oriented books. Regrettably, it has been allowed to go out of print.

Schuchert, Charles. 1943. *Stratigraphy of the Eastern and Central United States*. New York: John Wiley and Sons.

Scientific Literacy Workshop Field Trip Guide. 1993. Springfield: Illinois State Museum.

Shelton, John S. 1966. *Geology Illustrated*. San Francisco: W. H. Freeman. This classic collection of diagrams and duotone photos illustrates important landforms and geologic processes.

Skadden, Bill. 1978. *The Geology of Door County: A Self Guided Tour*. Sturgeon Bay, Wisc.: Golden Glow.

Thornbury, William D. 1954. *Principles of Geomorphology*. New York: John Wiley and Sons. A classic text from the generation preceding that of modern "process geomorphology." Quite accessible even to the nonspecialist.

————. 1965. *Regional Geomorphology of the United States*. New York: John Wiley and Sons. This classic text contains a good section on Illinois's physiographic provinces, and also a discussion of the Midwest's pre-Pleistocene river systems.

Travis, Russell. 1955. *Classification of Rocks*. Quarterly of the Colorado School of Mines 50:1.

Treworgy, Colin, and Russell Jacobsen. 1979. Paleoenvironments and distribution of low-sulfur coal in Illinois. In *Compte Rendu* of the Ninth International Congress on Carboniferous Stratigraphy and Geology. Available as Illinois State Geological Survey Reprint 1986E.

Van Diver, Bradford. 1985. *Roadside Geology of New York*. Missoula, Mont.: Mountain Press.

————. 1988. *Imprints of Time: The Art of Geology*. Missoula, Mont.: Mountain Press. Lovely, artful color photos of landforms.

Westbroek, Peter. 1991. *Life as a Geological Force*. New York: W. W. Norton. Treatments of how geologic factors have influenced life on earth are common enough; this well-written and nontechnical book turns the tables and explains how organisms have affected grand geologic processes.

Wiggers, Raymond. 1993. *The Amateur Geologist: Explorations and Investigations.* New York: Franklin Watts. For ages 12 to adult. Facts on fossils, rocks, minerals, and the geologic time scale. Source book for field trips and school geology projects.

Willman, H. B., et al. 1967. *Geologic Map of Illinois.* Urbana: Illinois State Geological Survey. This large, gorgeous wall map reveals both the complexity and the grand patterns of the Prairie State's bedrock.

————. 1975. *Handbook of Illinois Stratigraphy.* Illinois State Geological Survey Bulletin 95. Urbana: Illinois State Geological Survey. A detailed look at each rock formation found in Illinois.

Relevant Natural-History Texts

Crum, Howard. 1988. *A Focus on Peatlands and Peat Mosses.* Ann Arbor: University of Michigan Press. While Illinois's peatland sites are not cited, this is a highly recommended general guide to the botany and ecology of bogs and fens.

Evers, Robert A., and Lawrence M. Page. 1977. *Some Unusual Natural Areas in Illinois.* Biological Notes No. 100. Urbana: Illinois Natural History Survey.

Mohlenbrock, Robert H. 1971. Bell Smith Springs. *Outdoor Illinois* October:19–26.

Thompson, Janette R. 1992. *Prairies, Forests, and Wetlands: The Restoration of Natural Landscape Communities in Iowa.* Iowa City: University of Iowa Press.

Cultural History and Architecture

Bach, Ira J., ed. *Chicago's Famous Buildings.* 3rd ed. Chicago: University of Chicago Press. One of two good pocket guides to Chicago's landmarks. This book is handier if somewhat less comprehensive than Sinkevitch's *AIA Guide to Chicago* (see below).

Cronon, William. 1991. *Nature's Metropolis: Chicago and the Great West.* New York: W. W. Norton.

Dickens, Charles. 1842. *American Notes for General Circulation.* New York: Penguin Books. A view of America and Americans you won't see on those top-of-the-hour blurbs on PBS. Get ready to encounter your nineteenth-century counterparts, gorging, swearing, boasting, and expectorating their way westward.

Durant, Will. 1953. *The Renaissance.* Vol. 5 of *The Story of Civilization.* New York: Simon and Schuster. This great account includes mention of Pope Julius II sending Michelangelo to Carrara, Italy, to obtain marble from the same quarry complex that would be depleted four and a half centuries later by the construction of Chicago's Amoco Building.

Ebner, Michael H. 1988. *Chicago's North Shore: A Suburban History.* Chicago: University of Chicago Press.

Franklin, Kay, and Norma Schaeffer. 1983. *Duel for the Dunes: Land Use Conflict on the Shores of Lake Michigan.* Urbana: University of Illinois Press. The story of the torturous, decades-long battle to save a portion of the magnificent Indiana Dunes region east of Chicago.

Howard, Robert P. 1972. *Illinois: A History of the Prairie State.* Grand Rapids, Mich: William B. Eerdmans Publishing. The state's history seen through the filter of finance and economics.

Molloy, Mary Alice. 1992. *Chicago since the Sears Tower: A Guide to New Downtown Buildings.* Rev. ed. Chicago: Inland Architect Press.

Pauketat, Timothy R. 1993. *A Guide to the Prehistory and Native Cultures of Southwestern Illinois and the Greater St. Louis Area.* Illinois Archaeological Education Series No. 2. Springfield: Illinois Historic Preservation Agency.

Peck, Ralph B. n. d. History of Building Foundations in Chicago. Engineering Experiment Station Bulletin Series No. 373. Urbana: University of Illinois Press.

Power, Richard Lyle. 1953. *Planting Cornbelt Culture: The Impress of the Upland Southerner and Yankee in the Old Northwest*. Indianapolis: Indiana Historical Society. A superb study of early settlement patterns and cultural attitudes in Illinois and adjoining states. Highly recommended.

Sinkevitch, Alice, ed. 1993. *AIA Guide to Chicago*. San Diego: Harcourt Brace. A thorough and engaging guide to hundreds of Chicago's most architecturally significant buildings.

Twain, Mark. 1883. *Life on the Mississippi*. New York: Bantam Books.

Zukowsky, John. 1987. *Chicago Architecture, 1872–1922: Birth of a Metropolis*. Munich: Prestel-Verlag. A superb picture-and-scholarly-text book, with a particularly good account of the Wrigley Building, by Sally Chappell.

Travel Guides and Atlases

Illinois Atlas and Gazetteer. 1991. Freeport, Me.: DeLorme Mapping. An excellent, large-format collection of maps. A must for the geological explorer.

Schnedler, Marcia. 1992. *Country Roads of Illinois*. 2nd ed. Castine, Me.: Country Roads Press. Despite a few incorrect geology "facts," this otherwise delightful book is so good as a generalist's guide to rural Illinois that these mistakes are quickly forgiven. Written with wit, a personal point of view, and the occasional superior pun.

⇀ Index ⇀

Abbott formation, 166–67
Adams County, 171, 175
Agassiz, Louis, 59
Alexander County, 271
Alleghenian mountain-building event, 32
alluvial fans, 142–43
Alto Pass, 230–31
AMAX Wabash Coal Mine, 216–19
American Bottoms, 222, 225
Amoco Building, Chicago, 95, 101–2
animals, ancient. *See* fossils
Annawan, 46, 48
Anvil Rock sandstone, 249–50
Apple River, 23–25
Apple River Canyon State Park, 20–25
aquifers, 143, 191–92
Archimedes, 154
architectural geology: of Batavia, 88; of downtown Chicago, 94–104; of Lockport, 108
Aspdin, Joseph, 43
Athens marble. *See* Sugar Run dolomite
Aurora, 84–85

Bailey limestone, 235, 237, 273
Bald Mound, 84–86
Baraboo, Wisconsin, 2
Baraboo quartzite, 172
barchan dunes, 49, 215
Batavia, 84–85, 87
Bath terrace, 160–61
Bay City, 262, 264–65
Bay Creek, 239, 241, 265–66
Baylis, 170–71, 173
Baylis formation, 172–75
Beach Park, 66, 68
beach ridges, 67–68, 72
Beall Woods State Park, 216–18, 220–21
Beardstown, 55, 157–58, 164, 194
Bedford limestone, 99, 102–3, 166
Belknap, 262, 267, 269
Bell Smith Springs, 239–43
Bernhardi, Reinhard, 59
Beverly, 170–73
Beverly-Baylis upland, 170–75

Big Muddy River, 235, 237
Big Sink, 252, 259
biocalcarenite, 104, 150
bitumen, 109, 121
Bixby, 222, 226
blackjack, 18
Blackwell Forest Preserve, 84, 89–90
Blodgett moraine, 74, 77, 79–80
Bloomington, 182–83, 186–88
Bloomington glacial advance, 45, 49, 158
Bloomington moraine, 46–49, 182, 184, 186, 188, 195
Bloomington Ridged Plain, 45, 182–83, 204
Bloomington spillway, Lake, 186
Blue Island, 112–17
Blue Mound kame, 190, 195
Bluff Coast of Lake Michigan, 75–78
bogs, 60, 64–65
Bonpas, ancient Lake, 220
Bonpas Creek, 219
Boyington, W. W., 96
Braidwood, 134–37
breccia, 153, 177
Brown County, 165
Brussels terrace, 178, 180–81
Buffalo Hart moraine, 182, 188
Buffalo Rock State Park, 52–56
Bumastus, 123
Bureau County, 37, 45
Burksville, 222, 229
Byron, 32–33, 36–37

Cache River, 265, 267–69, 273
Cache River State Natural Area, 262, 268–69
Cahokia alluvium, 142, 213
Cairo, 3, 7–8, 264, 267, 270–71, 274–76
Calamites, 137
Calhoun County, 175, 181
Calhoun peneplain, 17, 175, 180
Calumet Sag Channel, 106, 110
Calumet stage of Lake Chicago, 71, 81; shoreline of, 68
Calymene, 122–23
Cambrian period, 3, 245; Illinois rock dating to, 35, 41

Camelback kame, 61–63
Canadian Shield, 2, 40, 70
Canton, 158
Cap au Grès faulted flexure, 178–79
Carbide and Carbon Building, Chicago, 95, 101
carbonate rocks, 27, 29–30, 44, 97, 119, 150–51, 166; role in forming sinkholes and karst features, 27–28, 223. *See also* biocalcarenite; coquina; dolomite; limestone
Carboniferous period, 4, 165. *See also* Mississippian period; Pennsylvanian period
Carlock, 182, 186–87
Carpentersville, 85
Carrara marble, 101–2
Carroll County, 26–27
Cary moraine, 60
Caseyville cuesta, 244, 247, 258
Caseyville formation, 233, 240, 247–50, 266
Castle Rock State Park, 38–41
cat-box filler, 8, 271–72
Catlin, 196, 201
Cave in Rock (town), 252, 254, 259
Cave-in-Rock State Park, 252, 259–61
caves in karst areas, 229, 259–61; epigenic and hypogenic types, 224; phreatic and vadose theories about, 224, 261
cement, constituents and production of, 43–44
Cenozoic era, 8–10. *See also* Tertiary period; Quaternary period
Cerro Gordo moraine, 190, 194–95
Champaign, 193
Champaign moraine, 194
channelization, 49–50, 267–69
channel lag, 167, 172
Chatsworth moraine, 128
Chazy group strata, 101
Chenoa, 182, 185
chert, 108–9, 121, 150, 259; origin of, 121
Chesterian series strata, 233, 265
Chicago: downtown area, 94–104; previous shorelines of, 94–95; unconsolidated lakebed sediments of, 94, 103–4
Chicago, ancient Lake, 56, 71, 79–81, 90, 93–94, 107, 109–10, 112–17
Chicago-Calumet Lacustrine Plain, 82
Chicago Cultural Center, 95, 102–3

Chicago River, North Branch of, 74, 77, 80; West Fork of, 74
Chicago Sanitary and Ship Canal, 80, 106–7, 109–10
Chicago Tribune Tower, 95, 98–99
Chicago Water Tower, 94–96
Cincinnati arch, 193
circle-pivot irrigation, 50
Clear Creek formation, 274
coal, 168, 275; comparison of bituminous coal and anthracite, 197; discovery in North America, 197; mining of, 52–53, 137–39, 201, 217–18; origin of, 5–6, 197–200; outcrops of, 53–54, 169, 218
Coal Age. *See* Pennsylvanian period
coal balls, 136
Coal City, 134, 136
Coastal Plain Province, 267, 269
Colchester coal, 52–54, 137
Collins, Ed, 62
Columbia, 222, 229
conglomerate, 167
continental glaciers. *See* glaciers
Cook County, 75, 93, 113, 119
coquina, 99
coral, sunflower, 22–23
Cordaites, 137
Crawford County, 209, 211
Cretaceous period, 7–8, 171, 265, 271–72
crevice deposits, 18
crevice mine, 19
crinoids, 150
crossbedding, 40, 225, 242, 266
Crown Point limestone, 101
crude oil. *See* petroleum
crushed stone, production of, 4
crust of the Earth, origin of, 1
cuestas, 247
Cumberland River, 129; course and diversion of ancient, 263–65
cyclothems, 5, 198–201

Dalton City, 203
Danvers, 182, 186
Danville, 196, 198, 201
Davis, William Morris, 30, 145
Dead River, 67–68, 72–73
Decatur, 194
Deerfield moraine, 74, 77, 80
Deer Plain terrace, 178, 180–81

Delavan, 157
Delavan till, 187
Dellwood Park, Lockport, 106, 108–9
Des Plaines channel sluiceway, 114
Des Plaines Disturbance, 74, 82–83
Des Plaines River, 74, 80, 106, 109, 134, 136
Devil's Backbone, 4–5, 233–35
Devonian period, 4, 178, 231, 250, 255–56, 273
Dickens, Charles, 271, 276
dikes, 256
dinosaurs, 7
disconformities, 133, 167
ditching, 49–50, 267–69
Dixon, 36–38, 43, 48, 50
Dixon-Marquette cement plant, 38, 43
Dixon Springs State Park, 262, 266
Dodgeville peneplain, 15–16, 19
dog mine, 201
Dolomieu, Guy de, 27
dolomite, 108, 121, 125, 150–51; definition of, 27; origin of, 29–30, 225
Door County, Wisconsin, 70, 109
Douglas, Paul, 83
Drake, Edwin, 209
driftless areas, 14, 21, 31, 175, 180, 204
Driftless Section, 14, 21, 31
Drury shale, 240
Dubuque, 30–31
Dundee, 85
dunes, 45–46, 48–49, 68–69, 82, 112, 116, 156, 159–61, 208, 215–16, 218
Dunleith formation, 22, 25
Du Page County, 85, 89, 91
Dupo, 222–23, 225

Eagle Valley syncline, 244, 249–50
earthquakes, 246–47
East Dubuque, 14
Easton, 158
East Saint Louis, 3, 176, 183, 222–23
Eau Claire shale, 186
Eddyville, 238
Elburn moraine complex, 84–85
Elco, 270, 273
Elgin, 85
Elizabeth, 16
Elizabethtown, 252, 254, 259
Elkhart, 182, 188
Elmhurst, 85

El Paso moraine, 182, 186
Embarras River, 213
end moraines. See moraines
Eocene epoch, 8, 272
Eospirifer, 123
Equality, 244
erosion-control barriers, 76–77
erratics, 35, 99
eskers, 42, 46–47, 86–87
Eureka moraine, 182
Evanston, 74–75, 82

Falling Spring site, 222, 224, 229
faults, 32, 41, 178–79, 230–32, 245–46, 266
Favosites, 122–23
fens, 60, 64–65
Fermilab National Accelerator, 88
ferns, seed, 6
flat-and-pitch deposits, 18
Fletchers moraine, 182
Flood of 1993, 143, 161, 235
floodplains, 161–62, 184
floods, ancient, 219, 221
fluorite, 253, 257–58
fluorspar, 253, 257–58
Forest Glen County Preserve, 196, 198–201
Fort Defiance Park, Cairo, 270, 275–76
Fort Payne formation, 255
Fort Wayne, 220
fossils: algae, 22–23; animals, 6–7, 29, 90–91, 120–23, 136, 154, 169, 272; plants, 6–7, 56–57, 135–37, 168–69, 271–72; reefs, 4, 119–25
Fountain Bluff, 230, 232–33
Fox.Lake, 58, 64
Fox Lake moraine, 60
Fox River valley, 84, 87–88
Fraction Run, 106, 108–9
Franconia formation, 35
French Canyon, Starved Rock State Park, 57
Frog City, 270, 275
Fuller's earth, 271
Fulton, 31, 141

Galena (city), 4, 13–19, 30–31, 34
galena (mineral). See lead ore
Galena group strata, 18–19, 22, 25
Galesburg, 144
Galesburg Plain, 145–47, 162, 165
Gallatin County, 245

Garden of the Gods, Shawnee National
    Forest, 6, 244, 247–48
Gardner, 135
Geode Park, Warsaw, 11, 148, 153
geodes, 154, 167
Gillson Park, Wilmette, 74, 82
glacial erratics, 35, 99
Glacial Park, 59–64
glaciers: characteristics of, 184; extent in
    Illinois, 9; origin of, 10, 183–84;
    thickness of, 188. See also specific
    glaciation/stage
Glasford, 83, 156
Glasford impact structure, 82, 156, 159
Glencoe, 78
Glenwood stages of Lake Chicago, 71, 79,
    81; shoreline of, 68, 115
gob piles. See spoil piles
Goethe, Johann Wolfgang von, 59
Golconda, 262–66
Goofy Ridge, 156, 161
Goose Lake State Natural Area, 134, 136
grabens, 245–46, 258, 266
Grafton, 176
Grand Canyon of Winnetka, 74, 78
Grand Detour, 33, 36, 38, 42
Grand Detour esker, 38, 42
Grand Tower, 230–31, 233, 235
Grand Tower formation, 234, 255
granite, 96–98, 101
Grassy Knob chert, 235, 237
Grayslake peat, 79
Grayville, 216, 219
Great Flood of 1993, 143, 161, 235
Green River Lowland, 45, 48–50
Grindstaff sandstone, 249–50
groins, 76–77
ground moraines, 184–85
Grundy County, 141
Gulf Stream, effect on
    Pleistocene climate, 10
gypsum, 153

Hajic, Edwin, 56, 160
Halysites, 122–23
Hamilton, 148, 150
Hamletsburg, 262–65
Hancock County, 149
hanging tributaries, 56–57, 131
Hanover, 14

Hansel, Ardith, 65
Hardin County, 253–61
Harrisburg coal, 218
Harvard, 58, 65
Havana, 146, 156–59, 161
Havana Lowland, 157–62
Havana terrace, 160
Haw Creek, Knox County, 146–47
Hebron, 58, 65
Hegeler, 196
Hennepin, 141
Hennepin Canal, 46
Henry, 134, 140, 142–43
Henry County, 45
Herod, 244, 252
Heron Pond Swamp, 268–69
Herrin coal, 199–200
Hicks Dome, 252–256
Highland Park, 74, 77–79
Highland Park moraine, 74, 77–81
Highwood, 74, 77
Hillsboro, 206
Hitchcock, Edward, 177
hogbacks, 233–34
Holocene epoch, 8
Hooppole, 46, 48
Horberg, Leland, 16
Horseshoe Lake, 7, 270, 274–75
Hubbert, Marion King, 209
Huntley, 58, 63, 65
Hutton, James, 51, 127, 131

Ice Age, 8
ice sheets. See glaciers
igneous rocks, 35, 256
Illinoian glaciation, 8–10, 25, 141, 145,
    147, 157, 188, 194, 203–4, 206, 219,
    233
Illinois and Michigan Canal, 52–53, 106–9
Illinois Basin, 2–3, 5, 37, 151, 209–11
Illinois Beach State Park, 66–73
Illinois Caverns State Natural Area, 222,
    228–29
Illinois Furnace, 252, 256–57
Illinois River, 80, 136, 141–43, 158–59,
    161–62, 181; origin of, 55–56
Illinois State Geological Survey, 93; public
    tours offered by, 11, 85, 89, 161;
    research conducted by, 32, 119, 153, 171
impact structures, 82–83, 156, 159

Impromptu Exploration Company, 209
Inland Steel Building, Chicago, 95, 104
interglacial stages, 8
Interlobate complex, 202, 204–6
Iowa River, ancient, 148, 150
iron production, 256
ironstone concretions, 135–36, 139, 201
isostatic rebound, 72
Isua series strata, 2

Jackson County, 4, 5, 230–32
Jasper County, 209, 211
Jersey County, 177, 181
Jo Daviess County, 4, 13–19, 21, 178
Johnson, W. Hilton, 65
Johnson County, 263
Johnsons Mound Forest Preserve, 84–86
Joliet, 96, 106, 128, 134, 183
Joliet formation, 130
Joliet marble. See Sugar Run dolomite
Jolliet, Louis, 197
Jordan Hill, 208, 213–14
Jurassic period, 7, 171, 246

kames, 48, 60–63, 85–86, 195, 202, 205–6
Kane County, 85
Kaneville esker, 84, 86–87
Kankakee, 126, 128
Kankakee County, 127
Kankakee formation, 132–33
Kankakee River, 129, 136
Kankakee River State Park, 126–33
Kankakee Torrent, 55–56, 127–30, 143,
    158–59, 185, 220, 233, 263, 265
Kansan glaciation, 8–9
Karnak, 262, 267
karst features and terrain, 28, 222–25,
    227–28, 252, 259–61
Keensburg, 216–17
Kentland impact structure, 83
Keokuk, Iowa, 148, 150
Keokuk limestone, 150–51
kettles, 60, 63–64
Kikidium, 123
Kimmswick limestone, 178–79, 225
Kingston, 170, 173–74
klintar, 123
knob-and-kettle terrain, 47, 65
Knox County, 145
Knoxville, 144, 146–47

Lake Bluff, 74–76
Lake Border Moraines Bluff Coast, 75–78
Lake Border morainic system, 74, 79–80,
    113–14
Lake County, 67, 75
Lake Forest, 74, 77, 79
La Moine River, 165
Lancaster peneplain, 16–17, 19, 30
landfills, sanitary, 89–90
LaRue Swamp, 230, 235–36
La Salle, 52, 134
La Salle anticlinorium, 43, 52, 54, 179
La Salle County, 51–52, 137
Lawrence County, 211
leachate, 90
lead ore, 4, 14, 17–18, 22; mining history,
    17; origin of, 17–18
Lee County, 39, 45
Lemont, 96, 106–7, 110
Lepidodendron, 137
Le Roy, 182–83, 187
Le Roy moraine, 182
levees, 235
Lexington, 182, 186
liesegang rings, 247–48
life, origin of, 2
limestone, 5, 99, 101, 104, 150–51, 198–99,
    273; definition of, 27; origins of, 29–30,
    150–51, 225, 273. See also carbonate
    rocks
limonite, 256
Lincoln, 158, 182, 188
Lincoln anticline, 179
Lincoln Hills section of the
    Ozark Plateaus Province, 180
Lingle formation, 234, 255
lithification, 172
lithographic limestone, 152
Little Mound, 156, 160
liverworts, 56
Livingston County, 183
Lloyd Park, Winnetka, 74, 80–81
Lockport, 106–8, 110
Lodge Park, Piatt County, 190, 193–94
loess, 162, 172–73, 181, 215
Logan County, 157, 188
longshore currents, 68
Longwall District, 134, 137–39
Longwall mines, 137–39
Lowden State Park, 33–37

Lower Rapids Gorge of the Mississippi
River, 148, 150
Lyell, Charles, 51

Magnolia, 134, 140, 142
Mahomet, 192, 195
Mahomet River, 157–58, 191–94
Mahomet sand, 191–92
Main Consolidated Oil Field, 211
Manhattan moraine, 128
Manito terrace, 160–61
Maquoketa group strata, 28, 31–32, 131, 179
Maquon, 144, 146
Marathon oil refinery, Robinson, 208,
210–11
marble, 97, 101–2, 104; black, 101
Marcus formation, 28
Marengo moraine, 65
Marquette, Père Jacques, 197
Marseilles moraine, 128
Marshall County, 137–38, 141
Mason City, 158
Mason County, 157, 160
Masters, John, 172, 175
mastodons, 90–91
Matthiessen State Park, 4, 52, 54
mature streams, 146
Maumee, ancient Lake, 220
Maumee Flood, 219, 221
Mazon Creek area, 134–37
Mazon Creek Biota, 135–37
McCormick, 238
McHenry, 58
McHenry County, 58–65
McLean County, 145, 183
McNairy formation, 265, 274
Mesozoic era, 7–8, 171. See also Triassic
period; Jurassic period; Cretaceous period
meterorite impact structures, 82–83
Metropolis, 262, 264
Michigan, Lake, 10, 67–72, 76, 78, 81,
183, 185
Michigan basin, 120
Mikulic, Donald, 136
Milankovitch cycle, 10
Minonk, 134, 139
Minonk moraine, 182, 185–86
Minooka moraine, 84, 128
Mississippian period, 4–5, 149–52, 165–67,
255, 257

Mississippi Embayment, 30, 272, 274
Mississippi Palisades State Park, 4, 26–32
Mississippi River, 30–32, 80, 127, 129, 141,
149–51, 181, 234–35, 265, 274; course
and diversion of ancient, 45, 55, 158, 203,
267, 270; origin of, 30–31
Monroe County, 223–24
Monticello, 190–95
moraine, 47, 147; origin of, 10, 183–85
Moraine Hills State Park, 65
Moraine Park, city of Highland Park, 74, 77
Moraine View State Park, 187, 195
Mound, The, 156, 160–61
Mounds gravel, 265, 274
Mount Carroll, 34
Mount Hoy landfill, 84, 89–90
Mount Simon sandstone, 186
Mount Vernon, 210
mud balls, armored, 158

Naperville, 85
natural bridge, 239, 241–42
natural gas, 186, 209, 227
Nauvoo, 148–50
Nebraskan glaciation, 8–9
Neda oolite, 126, 131–33
Nelson, W. John, 231–32
New Albany shale, 185, 255
New Harmony, Indiana, 220
New Madrid earthquakes, 127, 246–47
Newton, 208, 213
Niagara Falls, 109
Niagaran Escarpment, 70
Niagaran series strata and rocks, 28–29, 88,
108–9, 119–21, 178; glacially transported
pebbles from, 70
Nipissing, ancient Lake, 56, 71, 93–94, 107
Nippersink Creek, 60–62
Normal, 182, 187
Normal moraine, 182
North Branch of Chicago River. See
Chicago River, North Branch of
Northern Illinois Gas Station No. 40, 186
North Point Marina, Winthrop Harbor,
66, 69
Northwestern University shoreline
addition, 74, 82–83
No. 2 coal, 52–54, 137
No. 5 coal, 218
No. 6 coal, 199–200

Oblong, 208, 210–11, 213
Oblong Oil Field Museum, 208, 213
Oblong oil pool, 208, 211
Ogle County, 33, 39, 43, 53
Oglesby, 52
Ohio, Illinois, 46, 48–49
Ohio River, 129, 253, 259, 261, 265;
    ancient, 262–67, 270, 273
oil. *See* petroleum
oil wells, 210–12, 221. *See also* petroleum
Olive Branch, 270, 273–74
Olmsted, 270–71, 273
Olympia Center, Chicago, 95–97
Oneida moraine, 144, 147
oolite, 131–33, 225
ophiolite sequence, 98
Orange Township, 146
Ordovician period, 3–4, 40–41, 56–57, 133
Oregon, Illinois, 33–37
Oregon anticline, 41
Ottawa, 52, 54, 128, 134, 197
outwash, 48, 143, 159, 161, 213, 215
outwash plains, 48
Owen, David Dale, 220
oxbow lakes, 274–75
Ozark-Mahoning Company, 256

paleobotany, 136
Paleocene epoch, 8, 271–72
paleoecology, 119
Paleozoic era, 2–6. *See also* Cambrian;
    Ordovician; Silurian; Devonian;
    Mississippian
Pana, 202, 204, 206
Pangaea, 6, 137, 200
parabolic dunes, 49
Parkland sand, 116, 160
Park Ridge moraine, 74, 80, 112–13
Pascola arch, 210
Paw Paw Valley, 45
peatlands, 60, 64–65
Pecopteris, 137
Pekin, 141, 156–60
peneplain, origin of, 14–15
Pennsylvanian period, 4–6, 136–37, 165,
    167, 169, 197–201, 242
Pentamerus, 29
Peoria, 83, 141, 143, 158
Pere Marquette State Park, 176–81
Permian period, 6, 8, 257

Perrot, Nicolas, 17
Petersburg, 158
petroleum, 4–5, 212, 227; pay zones
    containing, 212, 227–28; production of,
    209–13, 221, 225
Piatt County, 191, 194
Pike County, 171
Pine Hills area of Shawnee National
    Forest, 230, 237
Pine Hills Escarpment, 4, 235–36
plants, ancient. *See* fossils
plate tectonics, 29, 151, 245–46
Platteville group strata, 34
Playfair, John, 51
Pleistocene epoch, 8–10, 149–50; drift
    exposures dating to, 187; extinction of
    mammals during, 90–91; tundra and
    taiga environments of, 90–91. *See also*
    Pre-Illinoian glaciation; Illinoian
    glaciation; Wisconsinan glaciation
Plum River fault zone, 32
point bars, 234
Pomona fault, 230–31
Pontiac, 134, 182–85
Pontiac, ancient Lake, 182, 185
Pope County, 239, 263
Porters Creek formation, 271–72
Portland Cement, 44
Potosi formation, 35
Pounds sandstone, 240–43, 247, 249
Powell, John Wesley, 30
prairies, 195
Pre-Illinoian glaciations, 9, 150, 157, 204
Precambrian basement, 2, 37
Precambrian rocks, 35, 70–71, 253
Precambrian time, 1–2, 8, 245
Princeton, 36–37, 45–46, 48, 55
Providence moraine, 46
Pulaski County, 271
Putnam County, 137

quarrying of stone, 87–88, 93
Quarry Park, Batavia, 84, 87–88
Quaternary period, 8–10. *See also*
    Pleistocene epoch; Holocene epoch
quicksand, 72

Racine formation, 28–29
radiocarbon dating, 71
Radnor till, 187

Ramsay, 206
Rattlesnake Ferry fault zone, 230–33
ravines, 77–78
Receptaculites, 22–23
reefs, fossil. *See* fossils
Ripley, 164–66
ripple marks, 73, 234, 243
rivers, geologic age of, 30. *See also* streams
Robinson, 208–11
Rochelle, 36
Rock Creek Canyon, 126, 130–31
Rock Creek graben, 254, 257–58
Rockdale moraine, 128
Rockford, 3, 34, 36–37
Rock River, 36; course and diversion of ancient, 36–37, 45
Rock River Hill Country, 34
rock shelters, 240–42
Rogers Park, 67
Rose Hill, 208–9, 215
Rose Hill spit, 74, 82
Rosewood Park, 74, 77–78
Rosiclare, 252, 254, 257–58, 264
Rough Creek graben-Reelfoot rift, 2, 245–46
Roxana silt, 187
rubble bars, 126, 129–30

Sag channel sluiceway, 114
Saline County, 245
St. Charles moraine, 84
St. Clair County, 223
St. Louis, 31, 141, 224
St. Louis formation, 152, 177–79, 224–27, 259–61
St. Peter sandstone, 3–4, 39–41, 52–57; quarrying and industrial applications, 39, 54
Ste. Genevieve fault system, 231
Ste. Genevieve formation, 224–25
Salem formation, 104, 166–67
sand dunes. *See* dunes
Sand Ridge State Forest, 156, 160–61
sandstone, 151, 167–69, 198–99
Sandwich fault zone, 41
Sangamonian interglacial stage, 194
Sangamon River, 193–95
Sangamon soil, 187
Sankoty sand, 143
Savanna, 3–4, 26–27, 29, 32, 34
Savanna anticline, 32

Savanna terrace, 178, 180–81
Scales Mound, 12, 14–15
Schuchert, Charles, 29–30
Schuyler County, 165
sedimentary rocks, world's oldest, 2
serpentine, 97–98
Seventeenth Church of Christ Scientist, Chicago, 95, 100
shale, 150–51, 167, 198–99
Shawnee Hills section of the Interior Plateaus Province, 240, 245
Shawnee National Forest, 235
Shawneetown fault zone, 246, 250
Sheffield, 46, 48
Shelby County, 203
Shelbyville, 202–5
Shelbyville glacial advance, 45, 55, 141, 150, 158, 194, 219
Shelbyville moraine, 182, 202–5
Shirley moraine, 182
Shumard, Benjamin, 220
Shumard oak, 220–21
Sigillaria, 137
Silurian period, 4, 29, 119–20, 123
sinkholes, 27–28
Skokie, 74, 81
Skokie Lagoons Forest Preserve, 79
Skokie River, 77, 79–80
slack piles. *See* spoil piles
Sonora formation, 152
sphalerite. *See* zinc ore
spoil piles, 137–39, 196, 201
Spoon River, 144, 146, 162
Springfield, 3
Springfield Plain, 204
Starved Rock State Park, 52, 56–57
Steno, Nicolaus, 149
Stephenson County, 17
Sterling, 36
Stockton, 20, 24
stone quarrying and production, 87–88, 93
Stony Island, 118, 123
streams: aggrading, 142, 159; diversion or reversal of, 23–25, 36–37, 61; stages of, 77–78, 146; underfit, 61, 162
Streeter, George "Cap," 94
Streeterville district of Chicago, 94
strip mines, 52–53, 137, 201
Sugar Run dolomite and formation, 87–88, 96, 108

Sunrise Park, Lake Bluff, 74–76
swell-and-swale terrain, 47, 65

Taconian Mountains, 120
Talbot, 156, 159
Tazewell County, 157
Teays River, 157–58, 191–94
Tennessee River, course and diversion of
    ancient, 264
terraces, 37, 143, 160, 178, 219–20
terra-cotta, 99
Terra Museum of American Art, Chicago,
    95, 97
Tertiary period, 8, 271–72. *See also*
    Paleocene epoch; Eocene epoch
Thebes, 263–64, 270
Thornton Quarry, 4, 109, 118, 121–25
till, 47, 63
till balls, armored, 158
till plains, 146, 184, 195, 204; origin of, 10
Tinley end moraine, 80, 106–7, 110, 184
Tinley ground moraine, 112–14
Toleston stage of Lake Chicago/Nipissing,
    68, 71
Towanda, 182, 186
Tower Beach, Winnetka, 74, 78, 80
Tower Hill, 202, 204–6
Tower Rock, 233–34
Tradewater formation, 166–67
travertine, 100
Treworgy, Colin, 200
Triassic period, 7, 171
Tripoli, 273–74
Trowbridge, A. C., 16
tufa, 87, 225
Tully Monster, 136

Udden, J. A., 198
underclay, 198–99
uniformitarianism, 51
University of Chicago, 119
Urbana, 193
Utica, 52

valley train, 37, 48, 220
Valparaiso morainic system, 80, 106–7,
    110, 114
Varna moraine, 140, 142
Velikovsky, Immanuel, 209

Vermilion County, 197
Vienna, 262, 266
Vinegar Hill Mine, 12
Volo Bog State Natural Area, 64–65

Wabash County, 217
Wabash River, 217, 220
Wade Township, 213
Walden Pond, Massachusetts, 64
Walker Hill, 233
Wanless, Harold Rollin, 198
Warsaw, 11, 148–50, 153
Warsaw formation, 152–54, 167
waterfalls. *See* hanging tributaries
Waterloo, 222, 229
Waterloo-Dupo anticline, 222, 226–27
wave-cut cliffs, 81, 112, 115
Weller, Marvin, 198
Wenona, 134, 138–39
wetlands, 79–80, 130–31
Whiteside County, 45, 141
Wildcat Hills, 244, 250
Will County, 107, 135, 137
Willow Springs, 110
Wilmette, 67, 74, 82
Wilmette spit, 74, 81–82, 114
Wilsonella, 123
Winnetka, 74, 76, 78–81
Winthrop Harbor, 66, 69
Wisconsin arch, 31
Wisconsin Driftless Section, 14, 21, 31
Wisconsinan glaciation, 8–10, 45, 107, 110,
    115, 141, 150, 157–58, 183, 188, 203, 263
Woman's Athletic Club, Chicago, 95
Woodbine, 16
Woodford County, 135, 137
Woodstock, 58, 65
Woodstock moraine, 65
woolly mammoths, 90–91
Worthen, Amos, 153
Wrigley Building, Chicago, 95, 98–100
Wyanet, 46–48

xenoliths, 35

zinc ore, 4, 17–18, 22; mining history, 17;
    origin of, 17–18
Zion, 66–68
Zion Beach Ridge Plain, 75

— About the Author —

Raymond Wiggers, an Illinois native, holds a bachelor's degree in geology from Purdue University. He served as editor and curator of publications for the Illinois State Museum, as an environmental geologist for the Illinois Environmental Protection Agency, and as a National Park Service botanist and environmental educator. Wiggers has written several books, including *The Plant Explorer's Guide to New England* (Mountain Press, 1994) and *The Amateur Geologist* (Franklin Watts). Wiggers currently resides in Beach Park, Illinois.

We encourage you to patronize your local bookstore. Most stores will order any title they do not stock. You may also order directly from Mountain Press, using the order form provided below or by calling our toll-free, 24-hour number and using your VISA, MasterCard, Discover or American Express.

Some geology titles of interest:

_____GEOLOGY UNDERFOOT IN NORTHERN ARIZONA	18.00
_____GEOLOGY UNDERFOOT IN SOUTHERN CALIFORNIA	14.00
_____GEOLOGY UNDERFOOT IN DEATH VALLEY AND OWENS VALLEY	16.00
_____GEOLOGY UNDERFOOT IN ILLINOIS	18.00
_____GEOLOGY UNDERFOOT IN CENTRAL NEVADA	16.00
_____GEOLOGY UNDERFOOT IN SOUTHERN UTAH	18.00
_____ROADSIDE GEOLOGY OF ALASKA	18.00
_____ROADSIDE GEOLOGY OF ARIZONA	18.00
_____ROADSIDE GEOLOGY OF SOUTHERN BRITISH COLUMBIA CAN: $25.00 US: 20.00	
_____ROADSIDE GEOLOGY OF NORTHERN and CENTRAL CALIFORNIA	20.00
_____ROADSIDE GEOLOGY OF COLORADO, 2nd Edition	20.00
_____ROADSIDE GEOLOGY OF CONNECTICUT and RHODE ISLAND	26.00
_____ROADSIDE GEOLOGY OF FLORIDA	26.00
_____ROADSIDE GEOLOGY OF HAWAII	20.00
_____ROADSIDE GEOLOGY OF IDAHO	20.00
_____ROADSIDE GEOLOGY OF INDIANA	18.00
_____ROADSIDE GEOLOGY OF MAINE	18.00
_____ROADSIDE GEOLOGY OF MASSACHUSETTS	20.00
_____ROADSIDE GEOLOGY OF MONTANA	20.00
_____ROADSIDE GEOLOGY OF NEBRASKA	18.00
_____ROADSIDE GEOLOGY OF NEW MEXICO	18.00
_____ROADSIDE GEOLOGY OF NEW YORK	20.00
_____ROADSIDE GEOLOGY OF OHIO	24.00
_____ROADSIDE GEOLOGY OF OREGON	16.00
_____ROADSIDE GEOLOGY OF PENNSYLVANIA	20.00
_____ROADSIDE GEOLOGY OF SOUTH DAKOTA	20.00
_____ROADSIDE GEOLOGY OF TEXAS	20.00
_____ROADSIDE GEOLOGY OF UTAH	20.00
_____ROADSIDE GEOLOGY OF VERMONT & NEW HAMPSHIRE	14.00
_____ROADSIDE GEOLOGY OF VIRGINIA	16.00
_____ROADSIDE GEOLOGY OF WASHINGTON	18.00
_____ROADSIDE GEOLOGY OF WISCONSIN	20.00
_____ROADSIDE GEOLOGY OF WYOMING	18.00
_____ROADSIDE GEOLOGY OF THE YELLOWSTONE COUNTRY	12.00

Please include $3.50 for 1-4 books, $5.00 for 5 or more books to cover shipping and handling.

Send the books marked above. I enclose $_____

Name_____

Address _____

City/State/Zip _____

☐ Payment enclosed (check or money order in U.S. funds)

Bill my: ☐VISA   ☐ MasterCard ☐ Discover ☐American Express

Card No. _____ Expiration Date:_____

Security No._____Signature _____

MOUNTAIN PRESS PUBLISHING COMPANY
P.O. Box 2399 • Missoula, MT 59806 • Order Toll-Free 1-800-234-5308
E-mail: info@mtnpress.com • Web: www.mountain-press.com